Get Qualified: Portable Appliance Testing

T0231633

- This book contains essential advice and guidance for those thinking of starting out in the portable appliance testing industry.
- A detailed look at the subject of portable appliance testing (PAT), this book is the ideal accompaniment for those studying the City & Guild and EAL PAT courses.
- Theory and assessment covered in one volume, with advice, revision exercises and sample tests to aid exam preparation.
- Contains all the information required to qualify and begin testing portable appliances.

The *Get Qualified* series provides clear and concise guidance for people looking to work within the electrical industry. This book clearly explains the options available to those wishing to enter the portable appliance testing industry and supports the reader through the subject in a step-by-step manner. Most importantly, it covers the theory behind portable appliance testing as well as looking in detail at each exam learning outcome. There are also sections on exam preparation, revision exercises and sample questions.

Kevin Smith is a highly regarded electrical trainer, with many years of experience delivering training courses to companies large and small throughout the UK. In recent years, in addition to running his own consultancy company, Kevin has worked within the electrical test equipment sector, most notably developing and delivering product training on behalf of the well-known test equipment manufacturer Seaward.

Get Qualified: Portable Appliance Testing

Kevin Smith

Routledge
Taylor & Francis Group

LONDON AND NEW YORK

First published 2017
by Routledge
2 Park Square, Milton Park, Abingdon, Oxon OX14 4RN

and by Routledge
711 Third Avenue, New York, NY 10017

Routledge is an imprint of the Taylor & Francis Group, an informa business

British Library Cataloguing in Publication Data
A catalogue record for this book is available from the British Library

Library of Congress Cataloging in Publication Data
Names: Smith, Kevin, 1978- author.Title: Get qualified. Portable appliance testing / Kevin Smith.Description: Abingdon, Oxon : Routledge, 2017.Identifiers: LCCN 2016003841 | ISBN 9781138189553 (pbk. : alk. paper) | ISBN 9781315641522 (eBook)Subjects: LCSH: Electric apparatus and appliances--Testing--Examinations--Study guides. | Household appliances--Testing--Examinations--Study guides. | Power tools--Testing--Examinations--Study guides. | Electrical engineers--Certification.Classification: LCC TK401 .S54 2017 | DDC 621.31/042076--dc23LC record available at https://lccn.loc.gov/2016003841

ISBN: 978-1-138-18955-3 (pbk)
ISBN: 978-1-315-64152-2 (ebk)

Typeset in Kuenstler by
Servis Filmsetting Ltd, Stockport, Cheshire

Acknowledgements

Many of the images in this publication have been included with the kind permission of Seaward Electronic Limited. The author wishes to thank Seaward for their cooperation and support during the writing of this book.

Contents

CHAPTER 1

Introduction

Portable appliance testing (PAT) or, to use the correct title, In-service Inspection and Testing of Electrical Equipment, has been with us since its adoption in the early 1990s, following the introduction of the *Electricity at Work Regulations* in 1989, which added some legal weight to the common sense requirement to maintain electrical equipment in a safe condition.

Twenty-five years later, PAT is still a huge industry in the UK, worth millions of pounds each year. Considering the size of the industry it is surprising to learn that most customers and a worrying number of those actually carrying out PAT have very little understanding of the subject. As often happens with health and safety practices, they get distorted over time so myths and conventions, which have little legal basis, become widespread, and untangling fact from fiction becomes impossible for anyone who does not have a detailed knowledge of the subject.

In this book I aim to simplify this subject and bring it back to its common sense roots. I will look at what is required to meet your legal obligations and explain the options you have to 'Get Qualified' and undertake PAT in your own right. Throughout the book I will make frequent reference to statutory documents, like the *Electricity at Work Regulations 1989* and to other guidance documents, such as the *IET Code of Practice*. A list of these documents can be found in the Suggested Reading section at the back of this book; many are available as free downloads and I would strongly suggest you use them as an accompaniment while you read the rest of this book.

The ability to undertake PAT is not rocket science and is easily within the reach of most people. With the right training and experience you will very soon be saving and even making money by carrying out this activity yourself. You will also have the added peace of mind that this important safety task has been done correctly and in accordance with current industry best practice.

CHAPTER 2

What is PAT Testing?

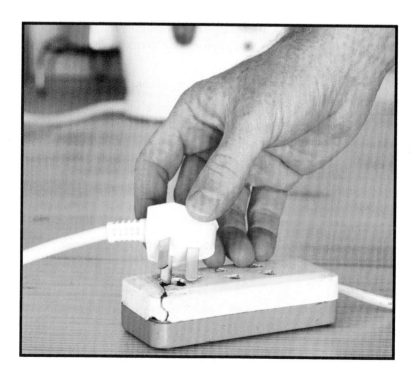

Most people will tell you that PAT testing is something you must have done every year and involves someone using a PAT tester to check the appliance and then fit a label to say it's safe for another 12 months.

The above is a common view but, as with most things, there is a bit more to it than meets the eye. Modern portable appliance testing is more of a system for managing the safety of electrical equipment while it is in service. One mistake that is often made is to focus too heavily on the 'testing' part of this system, which actually accounts for a very small percentage of the overall process and subsequently adds only a small amount to our knowledge of the equipment's safety status.

RISK

The foundation for all PAT testing should be a risk assessment, carried out by the person legally responsible for that equipment. This could be the owner or director of a company; however, in most cases this will be delegated to another person in the organisation, for example the health and safety manager. Risk assessments can be carried out individually for each item or generically for groups of items deemed to be similar in regards to the risk they represent.

When conducting the risk assessment we must consider many factors, including those listed below:

- The environment in which the equipment must operate.
- The type of people who may use the equipment.
- The equipment construction.
- The equipment type.
- How often the equipment is to be used.
- The way the equipment is installed.
- Any previous test records or information.

When the factors above have been established, the goal of the risk assessment is to utilise these factors to calculate the level of 'risk'.

FIGURE 2.1 How not to carry out a risk assessment

Risk is the combination of two judgements: **likelihood** and **consequence**.

Likelihood means, what is the chance that something will go wrong with this piece of equipment? Likelihood is affected by factors such as the type of equipment, where it's used and how often. A piece of hand-held equipment used every day on a construction site has a much greater likelihood of going wrong than a set of postal scales sat on the side in an office.

FIGURE 2.2   Risk is a balancing act

Consequence means, what is likely to happen if it does go wrong? Will the equipment simply stop functioning or could several people be killed?

In some systems actual numerical values are assigned to both likelihood and consequence; these are then multiplied together to get a 'risk score'. Often a threshold is set to determine which levels of risk are acceptable and which are not. If the level of risk is deemed to be unacceptable, control measures must be put in place to reduce the likelihood and/or consequences. The risk is then re-evaluated and the process continues until enough control measures have been put in place to reduce the risk to acceptable levels.

The in-service inspection and testing of electrical equipment (PAT) is a control measure and is only required when the person legally responsible for that equipment deems it necessary to reduce the risk posed by that equipment to acceptable levels. The nature and frequency of the in-service inspection and testing should also be determined as part of the risk assessment process and be specified as control measures that are proportionate to the risk.

TYPES OF PAT

There are three types of in-service inspection and testing detailed in the *IET Code of Practice*: user checks, formal visual inspections and combined inspection and test. A well-designed PAT testing system will utilise all three.

1. **User checks** are simple visual inspections carried out by the user of the equipment or another designated person (e.g. a school teacher or member of staff). These should be done either before use or at set daily or weekly intervals. The visual inspection is supplemented by a basic assessment by the user as to the equipment's suitability for the selected environment and the nature of task to be carried out. Users must be trained how to perform a user check and must be made aware of their responsibility to perform user checks at the intervals prescribed in the risk assessment. User checks are not recorded unless a fault is found and users must know the procedures for dealing with and reporting equipment believed to be faulty.

2. **Formal visual inspections** are always recorded and are carried out by 'competent' persons. They are very similar to the user check above, but will involve more detailed inspection in some areas, for example removing the plug top, where possible, to inspect its internal components and connections. Formal visual inspections are normally performed at longer intervals than user checks, often with several months between one formal visual inspection and the next.

3. **Combined inspection and testing** is the type of testing most people would identify as 'PAT testing'. In addition to the formal visual inspections carried out above, 'competent' persons also use electrical test equipment to perform a series of tests on the equipment to verify the serviceability of those components that cannot be checked by visual inspection alone. Again, this type of testing is usually carried out at greater intervals than the formal visual inspection, often with one or two years between tests.

The use of a PAT system that comprises all three types of testing is designed to give layered protection, which will reduce risk while keeping the amount of formal inspection and testing to a minimum, therefore reducing costs while maintaining safety.

Below is a chart (Figure 2.3) to illustrate the theoretical contribution each type of testing makes to identifying faults with electrical equipment.

The largest percentage of equipment faults should be identified during user checks, as these are carried out most frequently and users should be able to spot most major equipment defects during this type of inspection. Formal visual inspection comes next, as more detailed inspections

Where are most faults found?

- User checks
- Formal visual
- Combined inspection & testing
- Faults not discovered

FIGURE 2.3 The theoretical contribution each type of testing makes to identifying faults with electrical equipment

coupled with the 'competent' status of the inspector may lead to the discovery of items not picked up during the user checks.

Contrary to popular belief, combined inspection and testing makes the smallest contribution. This type of testing often only adds one or two electrical tests in addition to the inspection carried out as part of the formal visual inspection and it is rare for these tests to highlight a fault that was not already previously identified by other forms of testing.

By utilising a PAT system that takes advantage of the three types of testing, and coordinating the intervals at which these tests are carried out, we can put a control measure in place that identifies a majority of faults quickly, thus protecting the user, while reducing the need for frequent detailed testing, only required to identify a small number of rare faults.

Real life

Unfortunately the realities of PAT testing are somewhat different from the ideal scenario illustrated above. Most companies do not make use of user checks or formal visual inspections and opt simply for a combined inspection and test at a set interval, usually every 12 months. This system is less than optimal for several reasons:

- Any faulty equipment may remain in use for long periods of time between combined inspection and tests; during this time it poses a

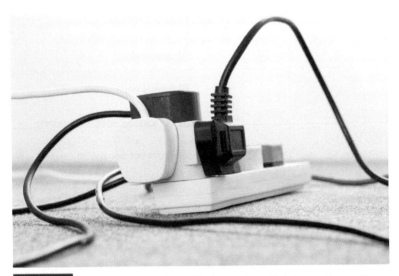

FIGURE 2.4   User checks can easily identify when equipment is being used incorrectly, for example this overloaded multi-way adaptor

significant risk to the user and the location as a whole. In addition to electric shock, risk of fire is one of the major hazards associated with electrical equipment.

- A rigid testing interval does not consider the individual levels of risk posed by each piece or type of equipment and therefore leads to many items being tested too much; and also, more worryingly, some items not being tested enough to control the level of risk they represent.
- Money is wasted on a system that has minimal impact on ensuring the safety of electrical equipment in the period between tests. Companies pay out large sums each year without knowing why and often get little more than new PAT labels in return.

A whole system view of PAT testing, starting with assessing the risk and then selecting suitable control measures is key to all good PAT testing.

CHAPTER 3

Who can PAT Test?

This question is the subject of much debate and hand wringing, but is essentially quite simple to answer. To carry out PAT testing you must be 'competent' to do so.

The person who decides if you are 'competent' or not is the person legally responsible for the electrical equipment and the premises in which it is to be used. Typically we are talking about company owners or directors who may delegate this decision to another person, for example a maintenance manager.

Establishing a person's competence is tricky and judgements must be made against two main criteria: **knowledge** and **experience**.

Get Qualified: Portable Appliance Testing. 978 1 138 18955 3 © Kevin Smith, 2017. Published by Taylor & Francis.

Regulation 16 of the *Electricity at Work Regulations 1989* (below) is often referred to when attempting to quantify competence.

> *No person shall be engaged in any work activity where technical knowledge or experience is necessary to prevent danger or, where appropriate, injury, unless he/she possesses such knowledge or experience, or is under such degree of supervision as may be appropriate having regard to the nature of the work.*

So the judgement is, has this person got enough knowledge and experience to avoid danger and injury, to themselves and others? And just as importantly could you prove it in court if it all goes wrong?

The *Electricity at Work Regulations* are not specific to PAT and therefore the above lacks detail: Knowledge of what? How much experience? Ultimately these are decisions for the person responsible, often referred to as the 'duty holder', but we can go some way to making that judgement easier. First, it is clear that there is no golden ticket that allows some to PAT test and others not. Proving competence is about compiling evidence of relevant knowledge and experience; the more evidence we have the better able we are to call ourselves 'competent' and the happier an employer will be to allow us to undertake work in their premises. Also remember that this is an ongoing process and as new guidance is published and new test equipment brought out, we must update our knowledge and experience to remain 'competent'.

CHAPTER 4

What are my Training Options?

Training is about building on the knowledge and experience you already have, therefore each person's training needs will be different. You must also consider the end goal: there is no point training to run a marathon if all you need to do is walk to the shops. Many people go for a course that is beyond their current ability level, subsequently falling at the first hurdle and spending a lot of money unnecessarily. It is always better to start small and build up as your knowledge and experience grow.

Get Qualified: Portable Appliance Testing. 978 1 138 18955 3 © Kevin Smith, 2017. Published by Taylor & Francis.

Essentially you will need to concentrate on the areas below:

Knowledge

- Electrical safety
- Electrical theory
- The *IET Code of Practice*

Experience

- Practical electrical work
- Using electrical test equipment
- Carrying out visual inspections
- Carrying out combined inspections and tests
- Completing records and recording test results

These areas form the basis of most PAT training courses and, depending on the depth of the course, they will cover each topic in more or less detail, with some courses concentrating heavily on one or two areas.

OPTION 1 – SELF TAUGHT

It is entirely possible to learn PAT testing on your own without attending a formal course. This route is particularly suited to those who already possess some knowledge and experience of the subject, for example an electrician or electronic engineer. Much of the information you need is available outside of a classroom environment, and with some background reading and practice you could soon be in a position to start PAT testing for real. The major downside of this option is that you will not receive any formal certification, which makes demonstrating your competence difficult. Many people following this route opt for taking their exams at a college or training centre on an 'exam only' basis, which can be a lot cheaper than attending a course if you shop around. If considering the self-taught route you must, however, be careful to work within your limitations and this route would not be suitable for those who are new to electricity and its dangers.

OPTION 2 – INFORMAL TRAINING COURSES

These types of course usually lasts from a few hours to a full day and will be delivered by a training company who have written the course content and will issue their own certificate to you on completion. This option is usually best for those who are new to the subject as generally the content is addressed in less depth and the trainer will be able to help you with any holes in your background knowledge.

When shopping for this type of course you must take great care to select a course that is right for you. Because the content of informal courses is set by the training company and not by a national body, the content and quality of these courses varies massively and so does the price, which can range from free to up to several hundred pounds. Generally you do get what you pay for, but be sure to ask plenty of questions before handing over your money.

Questions to ask before booking your informal course:

- How long is the course?
- What notes and handouts do you provide? (Get samples.)
- What is the expected trainer to delegate ratio?
- What is the background/experience of the trainer?
- How much practical content is there?
- Will we each have our own PAT tester?
- What method of assessment is used?
- Can I see a sample of the certificate I will receive?
- Do you have any reviews or testimonials I can see?
- What help/support is available after the course?

The better the answers you get to the above, the more you should be expecting to pay. Also, a good certificate from a well-known training company goes a lot further when proving your competence to a potential customer or employer.

Informal training courses often offer a better range of choice and have content you may not find in the more rigid structure of a formal course. This is especially the case when you are looking to get training on a specific type of PAT tester where the courses offered by the manufacturers are usually hard to beat.

OPTION 3 – FORMAL TRAINING COURSES

A formal training course is one in which the content and assessments are designed and regulated by a national organisation. These courses are generally longer than informal courses and go into more detail on the theory aspects. Formal courses have assessments that are set nationally and allow candidates to better prove their competence against national standards. The assessments usually consist of a formal practical assessment and a multiple-choice exam, both of which must be passed before candidates can receive their certificate from the awarding organisation.

There are two main types of formal training course available for portable appliance testing; these are offered by **City & Guilds** and **EAL** approved training providers. While City & Guilds is the most recognisable of the two, both hold equal standing within the industry and there are only minor differences between them.

The content for all formal PAT training courses is set nationally, with each organisation listing the specific criteria that must be achieved and the assessments that must be passed before a certificate will be issued.

Training providers wishing to offer these formal courses must first be approved to do so by the awarding organisation (City & Guilds or EAL) who will then carry out regular monitoring visits to ensure that the training provider operates to the high standards expected.

The two courses currently on offer are:

- City & Guilds level 3 awards in the *In-Service Inspection and Testing of Electrical Equipment (PAT)* and the *Management Requirements for the Maintenance of Electrical Equipment* (2377).
- EAL level 3 certificate in *In-Service Inspection and Testing of Electrical Equipment* (PAT) QCF – 600/4340/4.

When deciding whether to sign up, you must be aware that these formal courses are designed for those with an electrical background, and many people attending these courses without sufficient prior experience struggle to meet the necessary standard required to pass the assessments, even after several re-sits.

Formal courses are not for the faint hearted and I would only suggest you consider them once you have completed an informal course and gained experience in the field. That said, there is a lot we can do to improve your chances – with these courses, preparation is everything.

Formal course structure

Both the **City & Guilds** and **EAL** formal courses follow the same course structure. This is set nationally by OFQUAL, ensuring parity between awarding organisations. These formal courses are also heavily based on the content of the *IET Code of Practice*, which is used extensively throughout the courses and also during the exams. If you intend to undertake these formal courses you must have a copy of the latest version of the *IET Code of Practice*, which is also heavily referenced in this book.

The OFQUAL framework breaks the subject down into seven learning outcomes. A **learning outcome** is simply a goal that must have been achieved by the learner on completion of the course. Achievement of the learning outcomes must then be tested during formal course assessments. We can use the learning outcomes as a basic checklist to ensure that we have learned all the information necessary to pass both practical and theory exams.

The seven learning outcomes that make up the formal course structure for in-service inspection and testing of electrical equipment (PAT testing) are further broken down into **assessment criteria**. Each assessment criteria details a specific aspect of the learning outcome. So by mastering each of the topics listed in the assessment criteria we can master the subject as a whole.

In the next part of this book I will talk you through the seven learning outcomes and their associated assessment criteria. A key part of passing the multiple-choice exam, associated with the formal courses, is being able to relate the outcomes and criteria to the relevant information located in the *IET Code of Practice*. Often, you are not required to remember the information, simply just to know where to find it quickly under exam conditions. With this in mind, where possible, I will direct you to pages and chapters within the *IET Code of Practice* that cover the relevant criteria.

CHAPTER 5

Learning Outcome 1

Learning outcome 1

Understand the statutory and non-statutory requirements relevant to the management of electrical equipment maintenance

This learning outcome is specifically designed to test your knowledge of the various laws and documents relating to electrical equipment maintenance. There are a number of areas within the *IET Code of Practice* that provide information relating to this learning outcome.

Chapter 3, entitled 'The Law', which starts on p. 33 of the *IET Code of Practice* contains most of the information required to get to grips with this subject. It lists and gives some information on the key statutory documents relating to portable appliance testing and expands on areas such as scope, responsibility and maintenance.

I would also like to draw your attention to the appendices within the *IET Code of Practice*, specifically appendices II and III, which contain more legal references, and appendix IV, which lists the relevant guidance (non-statutory) material.

Assessment criteria 1.1

Specify the requirements of statutory and non-statutory acts and regulations in relation to maintenance of electrical equipment

The first and most important things to understand are the terms **statutory** and **non-statutory**. Statutory is the term used to describe something that is a legal requirement and therefore must be complied with by law. Something that is non-statutory, however, usually exists simply to provide guidance and is not a legal requirement. These definitions are a huge simplification, but will do fine for our purposes.

Typical examples of statutory documents include

- *Electricity at Work Regulations 1989* and
- *Health and Safety at Work Act 1974.*

While examples of non-statutory documents typically include

- *IET Code of Practice* itself,
- *BS 7671* (The IET wiring regulations) and
- health and safety executive guidance, such as *HSR25.*

Both statutory and non-statutory documents are important to us and it is essential that we know the difference between them.

Also see criteria 1.5 with regards to maintenance requirements.

Assessment criteria 1.2

State the scope of legislation with regard to the system voltage

This criteria is addressed in para. 3.2 on p. 37 of the *IET Code of Practice*.

This criteria is almost a trick question, because the legislation, specifically the *Electricity at Work Regulations 1989*, applies to all electrical 'systems'. The scope of legislation with regards to voltage simply applies to all voltages we are likely to encounter. In the UK the highest voltage used in our distribution network is 400,000 V or 400 kV, therefore this is listed in the *IET Code of Practice* as the scope for the legislation. So the voltages we could encounter range from battery powered equipment at just a few volts all the way up to 400,000 V. This is a typical exam question.

NOTE: Please do not confuse the above with the **scope of the code of practice**, which is addressed in criteria 1.7; this is a frequent mistake made during exams.

Assessment criteria 1.3

Identify premises to which the Acts and regulations apply

The *IET Code of Practice* touches on the subject of premises in para. 3.1 on p. 20; however, this is more in relation to the premises covered by the *IET Code of Practice* and not specifically those covered by the legislation.

In short, the *IET Code of Practice* is designed to cover the inspection and testing of electrical equipment in most types of premises. A couple of notable exceptions that must be taken into account are: the testing of medical equipment complying with *BS EN 62353*, which is outside the scope of the *IET Code of Practice*; also briefly mentioned in chapter 4.5 on p. 41 are premises where there is the potential for explosions such as petrol filling stations, which again fall outside of the scope.

Taking a wider view, questioning with regards to criteria 1.3 often relates to simply deciding whether equipment is being used *in* work premises or *for* work purposes. Typical examples may involve work equipment used by workers at work, work equipment used by workers working from home or equipment used at home by workers for personal DIY. You are often asked to decide which of these activities would fall inside or outside the scope of legislation. These questions can be tricky as there is no word-for-word answer to look up in the *IET Code of Practice*; you must instead rely on common sense and careful reading of the questions.

Assessment criteria 1.4

Identify the guidance given by the Health and Safety Executive relating to electrical equipment

The Health and Safety Executive (HSE) produce a wide range of guidance on all health and safety topics. Several key documents relate to portable appliance testing. A good summary of these documents is contained in appendix IV of the *IET Code of Practice*, which starts on p. 121.

The key HSE documents relating to portable appliance testing are:

- *HSR 25* – Memorandum of guidance on the *Electricity at Work Regulations 1989*. This document details and expands on each regulation, the meaning of key terms and suggested methods of complying with its requirements.
- *HSE guidance note GS 38* – Guidance on selection and use of test probes, leads, lamps, voltage-indicating devices and measuring equipment. This document gives advice on the safety of test equipment, particularly best practice relating to test leads and probes.
- *HSG107* – Maintaining portable and transportable electrical equipment. This HSE document is very similar to the *IET Code of Practice* and provides guidance on maintaining portable and transportable electrical equipment.
- *INDG236* – Maintaining portable electrical equipment in a low risk environments. This document gives guidance to those who carry out portable appliance testing in low risk environments such as offices.

These are the main HSE documents that relate to portable appliance testing; however, there are many additional documents referred to in appendix IV of the *IET Code of Practice*. It's a good idea to study these before your exam in order to have a good understanding of what guidance is available. Most HSE documents are available as free downloads from their website (www.hse.gov.uk); this makes them a great source of additional information.

I would also like to mention two further documents published by IET that you should be aware of:

- *BS 7671* – Requirements for electrical installations (The IET wiring regulations). Often referred to as the 17th edition wiring regulations,

this document is the standard used by electricians for all electrical installation work.
- IET guidance Note 3 – This document expands on the requirements of *BS 7671* with regards to inspection and testing of the fixed electrical installation.

Many questions relating to criteria 1.4 will simply ask which guidance document relates to which subject. Appendix IV of the *IET Code of Practice* is always a good starting point for this type of question.

Assessment criteria 1.5

Explain the legal requirement to maintain electrical equipment in a safe condition

This criteria specifically relates to maintenance and is covered in paragraph 3.4 on p. 38 of the *IET Code of Practice*.

The statutory requirement to maintain equipment stems from two main sources:

- Regulation 4(2) of the *Electricity at Work Regulations 1989* states that:
  As may be necessary to prevent danger, all systems shall be maintained so as to prevent, so far as is reasonably practicable, such danger.

  There are a few interesting terms used in the above that help us understand our maintenance requirements. The term 'as may be necessary to prevent danger' places a legal duty on us to establish the correct frequency for maintenance activities and is quite open-ended as long as we ensure that danger is prevented. The term 'all systems' refers to all electrical systems that could extend from simple battery-powered equipment up to a power station; naturally this would include the types of equipment on which we would perform PAT testing. The final term of interest is 'reasonably practicable'. This gives us the ability to apply a little common sense to the requirement – we must determine what is reasonable and practical. For example, we could PAT test each item ten times per day, but this would not be reasonable and practical so we must make a decision that balances the requirement to prevent danger against what is reasonable and practical in a real-world environment.

The legal requirement to maintain equipment in a safe condition reminds us of one of the main duties of the test operative during PAT testing, to identify if any maintenance is required. The test operative must look deeper than just physical faults or damage, but inspect wearing parts such as bearings, filters and brushes, for example, to see if maintenance is required.

- Regulation 5 of the *Provision and Use of Work Equipment Regulations 1998* states that:

  Every employer shall ensure that work equipment is maintained in an efficient state, in efficient working order and in good repair.

  Similar to the above requirement in the *Electricity at Work Regulations*, this requirement, however, uses the term 'efficient', which is used in relation to safety. Therefore, the above could be taken to say:

  *Every employer shall ensure that work equipment is maintained in a **safe** state, in **safe** working order and in good repair.*

  The meaning of the term 'efficient' in this context (i.e. safe) is often the basis for exam questions.

Assessment criteria 1.6

State the reasons for inspecting and testing electrical equipment and systems

As covered in criteria 1.5, the overarching requirement is to maintain equipment in a safe condition, therefore there is no specific imperative to carry out inspection and testing in their own right. An inspection and testing programme is therefore used to determine whether maintenance is required.

The *Provision and Use of Work Equipment Regulations*, specifically regulation 6 and the HSE guidance that accompanies these regulations, do however indicate that where the safety of work equipment depends upon installation conditions, and where conditions of work are liable to lead to deterioration, the equipment shall be inspected.

So the reason for inspecting and testing electrical equipment is to ensure that no maintenance is required, the equipment has not deteriorated to a point that would make it unsafe and, most importantly, that the equipment is safe for continued use.

Assessment criteria 1.7

Identify the scope of the code of practice for the in-service inspection and testing of electrical equipment

Chapter 1 of the *IET Code of Practice* is entitled 'Scope'. The term 'scope', often used in many such documents, means what is covered and what is not covered by this document. It is important when using any guidance material to ensure that the application with which you are concerned is within the 'scope' of that document.

The scope of the *IET Code of Practice* is broken down into several areas. These include: the persons for which the code of practice is intended; the range of equipment covered; the premises to which the code of practice applies; the range of voltages and phases; and finally a summary of the objectives of the code of practice.

One area within the scope often targeted during exam questions is para. 1.4 on p. 20. This paragraph relates to the standard voltages and phases to which the code of practice applies. Often, questions relate to the scope of the code of practice with regards to specific voltages. Many candidates get this confused with the scope of the legislation, which is covered in criteria 1.2.

The *IET Code of Practice* is designed for 'low voltage' equipment. The term 'low voltage' is often misunderstood and refers to a voltage of up to 1000 V AC or 1500 V DC between conductors, or 600 V AC or 900 V DC between conductors and earth. It also covers single, two and three phase equipment. Primarily, the *IET Code of Practice* is designed for the typical voltages that we are familiar with, specifically 400 V, 230 V and 110 V. Reference is also made to extra-low-voltage equipment, specifically SELV, also known as Class III equipment, which operates at voltages up to and including 50 V AC or 120 V DC between conductors or to earth.

CHAPTER 6

Learning Outcome 2

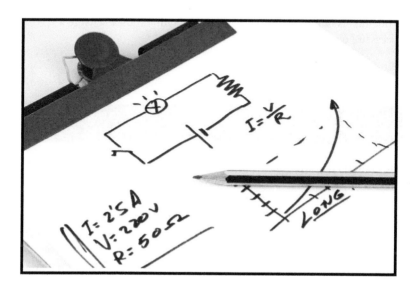

Learning outcome 2

Understand the electrical units of measurement associated with in-service inspection and testing of electrical equipment

For those without an electrical background, learning outcome 2 is a common source of problems, often contributing to poor exam performance.

Although this outcome specifically mentions *units*, candidates are generally expected to be able to carry out *calculations* using these units and this may involve applying simple equations such as Ohm's law.

Another factor that makes this outcome particularly tricky is the lack of information contained within the *IET Code of Practice*. Because this information is considered to be part of a person's prior knowledge, no content relating specifically to electrical units or theory is contained within the *IET Code of Practice*, so during the exam there is no way of simply looking up this information.

For the reasons mentioned above it is particularly important that before undertaking any formal exam for the subject you spend a lot of time practising questions involving units, electrical theory and calculations. While this is not a big part of any exam, it is important to ensure a successful outcome.

Assessment criteria 2.1

Identify the following SI units of measurement in relation to electrical equipment inspection and testing

This criteria requires you to have a good working knowledge of the standard units of measurement that you might come across during portable appliance testing. It is important that you not only know the units and the quantities they relate to, but that you have a good understanding of how their properties are affected, for example, with changing cable length or cross-sectional area.

Below is a list of the standard units relating to portable appliance testing. This is not an exhaustive list, but it does cover the main units and should form the basis for your revision.

- Ampere (Current)
- Volt (Voltage)
- Ohm (Resistance)
- Watt (Power)
- Hertz (Frequency)

CURRENT

Electricity is simply a flow of negatively charged particles, called *electrons*, from one place to another. Similar to the amount of water flowing in a river, the current in an electric circuit simply tells us how many electrons are flowing past a set point each second. The unit of current is the *ampere*, commonly known as the *amp*. The greater the flow of electrons, the more amps are said to be flowing in the circuit. Often the letter 'I' is used in equations to represent current.

All electrical equipment use a flow of electrons (a current) to perform its intended function, for example electrons flowing through the heating element within a kettle create heat, which is used to boil water.

A common example of the amp that you may be familiar with is the term '13 amp fuse'. Often, amps are expressed using the letter A, e.g. 13 A.

VOLTAGE

Negatively charged electrons are attracted to areas of positive charge: the greater the difference between negative and positive, the greater the energy with which the electrons move. The difference between positive and negative is often described as 'potential difference' and more commonly the 'voltage'.

The unit of voltage is the volt (V) and this is used to express the potential difference between positive and negative. For example, a battery with 12 V potential difference between its positive and negative terminals would commonly be described as a 12 V battery.

RESISTANCE

Each material has a unique atomic structure with varying numbers of electrons positioned at varying distances from the nucleus. Therefore the attractive force that binds the electrons to their parent atom varies greatly from material to material.

Some materials, such as copper, have only a very weak hold on some of their electrons; these materials are known as good conductors. Materials that are good conductors are said to have a low resistance, as they allow electrons and therefore electric current to move easily through them.

Other materials such as PVC or rubber hold tightly to their electrons and therefore would be described as good insulators. Insulators resist the flow of electrons and are therefore said to have a high resistance.

The unit of resistance is the ohm, which is often expressed using the symbol Ω.

As electricians we take advantage of the properties of materials, particularly conductors and insulators, to control the flow of electrons, making electricity travel where we want it to go.

Below is a simple diagram of a flexible mains cable:

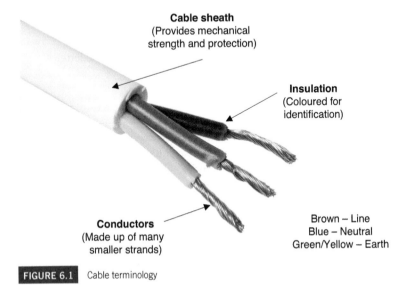

Cable sheath
(Provides mechanical
strength and protection)

Insulation
(Coloured for
identification)

Conductors
(Made up of many
smaller strands)

Brown – Line
Blue – Neutral
Green/Yellow – Earth

FIGURE 6.1   Cable terminology

Flexible cables are often classified by the size of their conductors, usually expressed in mm². This number relates to the cross-sectional area of each of the conductors, for example a 0.75 mm² three-core cable will have three conductors, each having the cross-sectional area of 0.75 mm². The cross-sectional area of a cable can be easily identified, as it is usually printed onto the sheath at regular intervals along its length at the time of manufacture.

Properties affecting the resistance of a conductor

The resistance of a conductor is dependent on four properties; you can remember these properties by using the acronym **MALT**.

Material

As I mentioned earlier, each material has a different atomic structure therefore each material will have a different resistance: this property is known as its *resistivity*. Usually this property has a limited effect during PAT testing as most conductors are made of standard materials, such as copper, and these materials are unlikely to change during the life of an appliance.

Area (cross-sectional area)

The cross-sectional area (CSA) of a conductor, when increased, allows electrons to flow more easily through it. This has the effect of lowering the conductor's resistance. The opposite effect can be observed when a conductor's cross-sectional area is reduced, thus increasing its resistance. The cross-sectional area of a conductor is therefore said to be inversely proportional to its resistance.

If the cross-sectional area of a conductor is doubled, its resistance halves. If we halve the cross-sectional area the resistance will double. This is often an area explored during exam questions, a typical example being:

You have a 0.75 mm^2 cable with a measured resistance of 0.5 ohms. If this cable were to be changed for a 1.5 mm^2 cable what would be the effect on the measured resistance?

If we double the CSA we halve the resistance, therefore the answer would be 0.25 ohms.

Length

The length of a conductor is often a key factor that must be considered when portable appliance testing, specifically in relation to testing long extension leads. If we increase the length of a conductor we increase its resistance and if we shorten a conductor its resistance will decrease. Therefore the length of a conductor and its resistance are said to be directly proportional.

If the length of a conductor doubles then its resistance will also double, if the length is halved then again the resistance will halve.

This property is also explored during exam questions, a typical example being:

You have a 10 m extension lead with the measured conductor resistance of 1 ohm. If the length of the lead is reduced to 5 m, what will the new measured resistance be?

In this case halving the cable length will halve the resistance so the correct answer would be 0.5 ohms.

Temperature

The temperature of a conductor also affects its resistance. If we increase the temperature of a conductor we will also increase its resistance; if we lower the temperature of a conductor we will decrease its resistance.

While changes in temperature do affect a conductor's resistance, temperature is not a property often taken into account during portable appliance testing. PAT testing is usually assumed to have taken place at a nominal ambient temperature of 20°C and any small changes in temperature above or below 20°C are not considered to significantly affect test results. However, if a high level of accuracy is required the test operative should take this into account and alter pass/fail limits accordingly.

Appendix VI of the *IET Code of Practice* contains table VI.1 on p. 134. This table will be used extensively during your exam so you should spend some time familiarising yourself with its layout and the values it contains. This table contains standard resistances per metre for a variety of different sized conductors at an ambient temperature of 20°C. We can use it to calculate conductor resistances for any length of cable. There is also additional information on maximum current-carrying capacity and cable diameter.

Ohm's law

Ohm's law is the equation that defines the relationship between current, voltage and resistance. When given any two of these three properties we can use Ohm's law to calculate the third.

Ohm's law can be expressed using a simple triangle, which is also a useful memory aid during your exam.

We can use this triangle to remember the three Ohm's law equations: you simply cover up the property you are trying to find with your finger, then the relationship of the other two properties gives you the other half of the equation.

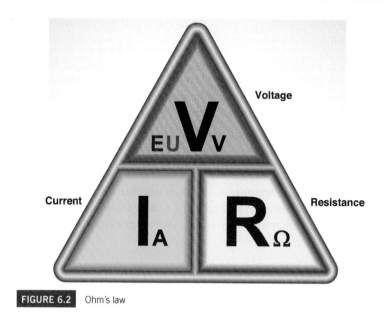

Ohm's law

For example: If we are trying to find voltage and already have the current and the resistance, we simply place a finger on the V for the voltage we are trying to find and read off the IR from the triangle.

This gives us the equation $V = IR$ or $V = I \times R$

We can continue this process for the other two units giving us the equations below:

$$I = \frac{V}{R} \text{ and } R = \frac{V}{I}$$

A simple example of an Ohm's law-based question is shown below:

The resistance of the heating element in a kettle is measured to be 23 ohms, the kettle is supplied from a mains voltage of 230 V, what current will flow in the element?

In this question we are given the resistance and the voltage, and from these values we are asked to calculate the current. Using the triangle, place a finger over the current, this tells us we are looking for V over R, or $V \div R$. So to get the correct answer we simply divide the voltage, 230 V, by the resistance, 23 ohms, which gives us an answer of 10 A.

The question above is typical of the type of question you could be facing during an exam, if you find this type of question difficult I would suggest practising until the calculations become second nature. This will not only help your confidence during the exam, but also increase your final mark.

POWER

Electrical power is the combination of voltage and current, which together give us the ability to do work. The unit of power is the watt, and everyone is familiar with terms like a 100 W light bulb, or a 10 kW shower.

Power can also be expressed in the form of an equation:

Power = Voltage × Current

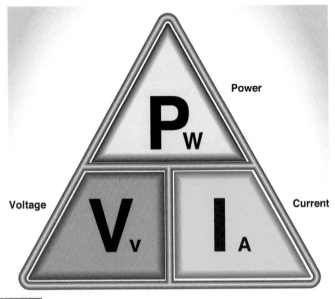

FIGURE 6.3 Power triangle

The power equation can be transposed using the above power triangle in the same way that we extrapolated the three equations relating to Ohm's law.

$$P = V \times I \qquad\qquad V = \frac{P}{I} \qquad\qquad I = \frac{P}{V}$$

The above equations can be very useful. The voltage is often a constant, i.e. 230 V, so these equations are often used to calculate the current when we know an appliance's power rating, or to calculate the power when we know the current.

Below is an example of how the power equation may be used during an exam question:

A mains powered, 230 V, kettle has a power rating of 2300 W or 2.3 kW. What current will flow when the kettle is switched on?

Using the power triangle, we place a finger over the I for current, which gives us P over V, or $P \div V$. Therefore, we must divide 2300 W by 230 V, which gives us an answer of 10 A.

This type of question does not occur frequently within exams, but again if this is an area you particularly struggle with, it's worth putting in a bit of time practising examples until you become comfortable with this style of question.

FREQUENCY

Most people have heard of the two types of electric current, DC (direct current) and AC (alternating current).

A typical example of DC is a simple 12 V battery. In a battery the positive terminal always remains positively charged and the negative terminal remains negatively charged, so the electrons always flow in one direction around the circuit, therefore the battery's output is said to be a constant 12 V DC.

Alternating current is most commonly found in the mains distribution network. The mains socket outlets in your home provide a nominal voltage of 230 V AC. Alternating current, as the name suggests, does not have a constant polarity: the flow of electrons first flows in one direction, then in the other. Our 230 V mains voltage actually changes direction

50 times per second; this is known as its frequency, which we state as 50 Hz. Due to the speed with which the current flow changes direction we usually do not notice the difference between direct current appliances and those that operate on alternating current.

It is not necessary, during your exam, to have a detailed knowledge of frequency. You simply need to know the difference between AC and DC and that the nominal mains supply voltage is 230 V AC with a frequency of 50 Hz.

Assessment criteria 2.2

Identify the multiples and sub multiples of SI units

We are all familiar with multiples and sub multiples used in everyday life, although we might not notice them. Typical examples include kilograms, millimetres, megahertz and gigabytes. One of the strengths of metric systems of measurement is that metric units can be easily expressed using multiples and sub multiples, which allows us to easily convert values into more meaningful numbers, giving us greater understanding and the ability to avoid the errors associated with long numbers containing lots of zeros.

All of the SI derived units described in criteria 2.1 can be expressed using multiples and sub multiples. Below is a table showing the typical range of values encountered during portable appliance testing:

Table 6.1 The typical units encountered during portable appliance testing

Value	Definition	Typical example
Mega (M)	1 million or 1,000,000 of something	$M\Omega$
Kilo (k)	1 thousand or 1,000 of something	kW
Unit	1 of something, i.e. the base unit	A, V, Ω, W, Hz
milli (m)	1 thousandth or 0.001 of something	mA, mΩ

Many problems with calculations stem from the incorrect conversion of multiples and sub multiples into the base unit. Prior to carrying out any calculation it is essential that all values are converted to the base unit, i.e. amps, volts, ohms etc.

There are many common strategies used to convert multiples and sub multiples to their base unit and ultimately whichever system you use the result should be the same. My advice, when it comes to portable appliance testing, is to keep it simple. So below is an example of a typical strategy I would recommend:

To convert from milli to units we simply divide by 1,000: for example, to convert 100m Ω into Ω, we divide 100 by 1,000, which gives an answer of 0.1 Ω.

To convert from kilo to units we simply multiply the value by 1,000; for example, 10 kW expressed in W is 10 multiplied by 1,000, or 10,000 W.

To convert from mega to units we simply multiply the value by 1,000,000, for example 2 M Ω expressed in Ω, is 2 multiplied by 1,000,000, or 2,000,000 Ω

The above conversions can be reversed by substituting multiply for divide and vice versa. I would recommend using a calculator to carry out all conversions. This may seem over the top, but experience has shown that in exam conditions people often make silly mistakes when doing these conversions in their head or on paper. It is important before your exam that you practise converting values, especially with regards to converting values in table VI.1 on p. 134 of the *IET Code of Practice*, which are expressed in mΩ, into Ω, by dividing these values by 1,000. During most exams you will be required to perform this exercise several times.

CHAPTER 7

Learning Outcome 3

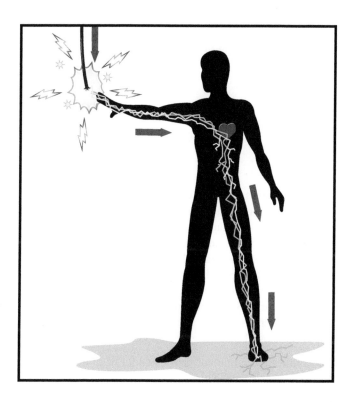

Learning outcome 3

Understand how equipment construction and classification reduces the risk of electric shock

This learning outcome requires that we learn a little about how electrical equipment and appliances are constructed. In addition to allowing the appliance to perform its intended function, its construction also protects us against electric shock.

In order to perform in-service inspection and testing on an item of equipment, we must first know how it is constructed and what measures the designer took to provide electric shock protection to the user. Once we know what protective measures are employed within the appliance, we will then know what items to inspect and what items to test. Identifying the protective measures within each different item of equipment encountered is one of the most difficult tasks faced by portable appliance testers and in this regard there is no substitute for experience.

Assessment criteria 3.1

Identify the types of electrical equipment

Although the law makes no such distinctions, the *IET Code of Practice* separates electrical equipment and appliances into 'types'. The type of equipment approximately relates to the risk it poses and therefore this impacts on frequency of tests and in some cases pass/fail limits. It is a common mistake to assume that PAT testing only applies to 'portable' equipment; in actual fact 'portable' is just one of the types described in the *IET Code of Practice*. It is common sense that appliances such as washing machines or cookers aren't exactly portable, therefore other types are defined that better describe the nature of the wide variety of equipment which will come across during in-service inspection and testing.

On first encountering equipment, it is usual for us to make a decision about the equipment's type. This is often recorded and used to form the basis for decisions made later on in the process. Due to the wide variety of equipment that may be encountered on a day-to-day basis, it is easy to see that not every piece will fit easily into a single type or definition, and disagreements over an equipment's type are to be expected. As the law simply refers to electrical 'systems' and not equipment types, the allocation of types is not a black-and-white process and should be made, using a competent person's knowledge and experience, on a best fit basis.

The *IET Code of Practice*, chapter 5, starting on p. 43, defines these 'types' of equipment. Below, I will elaborate on the definitions given, and attempt to clarify which equipment belongs to which 'type'.

PORTABLE

A key starting point for identifying equipment as 'portable' is that it must weigh less than 18 kg. Portable equipment is also usually intended to be moved while in operation or can be easily moved from one place to another. On first sight this definition appears straightforward, but difficulties arise due to significant crossover with other types, such as movable, hand-held and IT equipment. So an item such as a drill, which would, at first, appear to fit neatly into the 'portable' definition, is actually classified as 'hand-held'. Care must be taken when assigning an equipment's 'type' to ensure you have considered all the options and chosen the best fit. It is also worth pointing out that no mention is made in the definition to the appliance having a plug, or not. A common PAT testing myth is that a portable appliance is simply any item with a plug; this is not true and not referred to in either the *IET Code of Practice* or the law.

FIGURE 7.1   Portable appliance

FIGURE 7.2   Movable appliance

MOVABLE

Movable equipment is actually referred to as transportable equipment in some HSE documents; this fact is occasionally the subject of exam questions.

The definition for movable equipment is split into two parts, the first part being very similar to the definition for portable equipment: equipment that weighs 18 kg or less and is not fixed. The weight limit of this definition aligns with that of portable equipment, so the separating factor appears to be that these items are not designed to be moved whilst in operation and are not easy to move from one place to another. They are also not fixed.

As you can see, very little equipment would actually fit into this definition and, as a result, most equipment classified as 'movable' will actually come from the second part of the definition: equipment with wheels, castors or other means to facilitate movement by the operator. As no weight limit is mentioned it is assumed that equipment meeting the second part of the definition is not limited by its weight.

Therefore a simplification of this definition would be *equipment of any weight that is designed to be moved by the use of wheels etc.* This could include large industrial equipment such as welders and pressure washers. It is easy to see how confusion can arise due to crossovers with other types, especially portable equipment.

HAND-HELD

This type of equipment is designed to be held in the hand during normal use and would typically include items such as drills, hedge cutters, irons etc.

Identifying hand-held equipment is very important as this type of equipment is usually deemed to be high risk. Risk, as explained earlier, is a combination of likelihood and consequence. Hand-held equipment, due to the way it is used, has a greater likelihood of requiring maintenance. This could be due to damage to the cable, damage to the enclosure or simply frequent use. Also, hand-held equipment is designed to be held in the hand, therefore if a fault develops the consequences could be greater,

FIGURE 7.3 Hand-held appliance

due to the increased contact between the user and exposed parts, making an electric shock more likely.

The higher risk posed by hand-held equipment usually results in the item being subject to more frequent inspection and testing. In an effective PAT testing system particular attention should be paid to high-risk items, such as hand-held equipment, as strict management of these appliances will make a big contribution to the overall electrical safety.

STATIONARY

Equipment weighing more than 18 kg and not provided with a carrying handle is classified as 'stationary'. Typical items of stationary equipment include washing machines, dishwashers and refrigerators. Stationary equipment is usually considered to represent a relatively low risk as

FIGURE 7.4 Stationary appliance

items are generally connected to the same socket throughout their life, leads are kept out of harm's way behind the appliance, and contact with the user is limited to simply switching the device on or off. As a result, most items of stationary equipment are not subject to inspection and testing as frequently as other, high-risk, items.

FIXED

This type of equipment is either fixed in a specific location, i.e. screwed to a wall, or wired directly to the fixed electrical installation and not connected via a plug and socket arrangement. Typical examples of fixed equipment include hand dryers, heated towel rails and cookers.

It has been a common myth for many years that PAT testing does not apply to fixed items. I am even aware of many instances where companies removed an appliance's plug and hardwired the appliance directly into a fused connection unit to remove the necessity for PAT testing. As we learned earlier, our legal obligations make no mention of the method of connection and therefore by removing the plug in

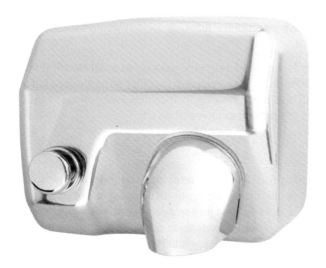

Fixed appliance

this way all that is achieved is to make the appliance more difficult and dangerous to test. While fixed equipment is generally considered low risk, due to its fixed nature and permanent supply connection, it requires careful isolation from the mains supply before inspection and testing can be carried out. The need for isolation means that the level of competence required to conduct inspection and testing on this type of equipment is greatly increased, making this type of inspection and testing a job for trained electricians, outside the competency range for most PAT testers.

Most companies have holes in their PAT testing system, caused by their failure to correctly manage the risks posed by fixed equipment. In many cases no inspection and testing is actually carried out on items such as hand dryers and cookers. The PAT tester believes that this type of equipment is not their job as it does not have a plug and the electrician who carries out periodic inspection and testing of the fixed electrical installation only tests up to the fused connection unit or cooker switch. Systems must be put in place so that all items are covered and no grey areas exist. The testing of fixed appliances throws up many dilemmas and knowing where to draw the line is a difficult decision for all duty holders. The *IET Code of Practice* lists among its examples of fixed equipment: central heating boilers, immersion heaters and air conditioning units, items that would have traditionally been ignored.

The duty holder must consider the risks posed by all electrical systems and put measures in place to ensure that they are maintained in a safe condition. These measures may include increased user checks and visual inspections to supplement electrical testing.

BUILT-IN

Generally recognised as a sub-type of fixed equipment, 'appliances or equipment for building in' relates to that small group of equipment designed to be installed in a prepared recess, such as a cupboard. A built-in electric cooker for example may not have the same protection around its supply terminals as would be the case for a freestanding unit, because some of the protection would be provided by the cupboard in which the cooker was installed, thus limiting access to live parts. Great care must be taken when testing built-in equipment, and visual inspection may require dismantling the equipment to some degree.

FIGURE 7.6 Built-in appliance

Again a high level of knowledge and experience are required for those carrying out this type of inspection and testing.

INFORMATION TECHNOLOGY (IT) EQUIPMENT

The term 'IT equipment' is now quite dated and can be misleading when used with regards to PAT testing. IT equipment was originally used to define equipment containing sensitive electronic components and circuitry; typically at the time this meant computers, monitors, fax machines etc. However in the modern world even a toaster or kettle may now contain electronic components, microcontrollers and LCD displays. So IT equipment as a type should be taken to include any equipment with sensitive electronic components, which is significantly wider than the original definition. This type of equipment is usually taken to be low risk.

The reason IT equipment is defined as a separate 'type' is due to the fact that equipment containing sensitive electronic components could be subject to damage during combined inspection and test. It is therefore essential that this type of equipment is identified prior to test and that the correct precautions are taken during testing to ensure that no damage is caused. Specifically, care must be taken when carrying

FIGURE 7.7 IT equipment

out earth continuity testing using high test currents and insulation resistance testing at 500 V DC.

OTHER SUB-TYPES

- Extension leads and RCD extension leads.
- Multiple adapters and RCD adapters.
- Surge protective devices.

The above are all listed in the *IET Code of Practice* but are generally not considered to be 'types' in their own right. I have referred to these as sub-types because they all generally fall within the definition of portable equipment. The *IET Code of Practice* recommends that leads, adapters and extension leads, where detachable from the appliance, be treated as separate items, recorded individually on the asset register, and inspected, tested and labelled as appliances in their own right.

Exam questions relating to this assessment criteria will normally describe an item of equipment mentioning its weight and a few other clues, which can be used to narrow down the correct answer. For example:

A 20 kg washing machine connected to a socket outlet in a kitchen would be classified as?

In this case, the item is over 18 kg, no mention is made of a carrying handle and the item is connected to a socket outlet, therefore not fixed. The combination of these factors would lead us to believe that the washing machine would be 'stationary' equipment. To aid in making this type of question as clear as possible, examiners tend to stick to the examples listed in the *IET Code of Practice*, so look out for this during your exam.

Assessment criteria 3.2

State the classification of equipment construction

In addition to identifying equipment by type, we also identify equipment by its class of construction. In very simple terms there are five different classes of equipment ranging from class 0 to class III and the higher the class number the lower the risk it poses to the user.

When a designer designs a new appliance they must consider how the user is to be protected against electric shock. There are many ways in which this can be achieved, so to make things simpler, the designer will choose one of the five standard methods, or classes.

The *IET Code of Practice* chapter 11, starting on p. 71, details the standard equipment classes used to classify portable appliances. It is essential when carrying out portable appliance testing that we correctly identify to which class a particular item belongs. Once we know what class an appliance is, we then know how the user is protected against electric shock and therefore which items need to be inspected and tested to confirm that the appliance is safe for continued use.

Today, a vast majority of appliances will either be class I or class II, the other classes are very rare, so have been included only for completeness. All classes are, however, mentioned in the *IET Code of Practice* so should form the basis of your exam revision.

When considering classes, I find it useful to look at them as a chronological progression, marking the evolution of electrical safety through appliance design. Below I will attempt to expand on the descriptions given in the *IET Code of Practice*.

CLASS 0

A class 0 appliance is one that has basic protection only. This type of appliance has no fault protection at all, so if the basic insulation should fail the user is immediately placed at risk. To find an example of this type of equipment, we can think back to the earliest electrical appliances.

Consider an old brass electric lamp. This lamp would be supplied via a round two-pin plug and have a two-core rubber insulated flex. Protection against electric shock was provided by the rubber insulation; if the insulation were to fail the body of the lamp would become live. The lack of any fault protection places the user at immediate risk. At the time, these levels of protection were considered adequate, but now such class 0 appliances are not considered safe for normal use and would warrant an instant failure during any visual inspection.

The most common example of class 0 equipment that you may come across during portable appliance testing are the old, mains powered Christmas lights. A two-core cable without sheath is connected into a standard 13 amp plug, there are no earth connections and protection against electric shock is provided by basic insulation alone. The lack

FIGURE 7.8 Old class 0 Christmas lights

of any fault protection makes this type of appliance unsuitable for continued use.

CLASS 0I

This is a pseudo-class, being a hybrid of class 0 and class I. This equipment is designed and constructed as class I equipment, it is, however, fitted with a two-core supply cable and two-pin plug, which has the effect of removing the connection between exposed metal parts and the earth in the fixed electrical installation. By circumventing the fault protection in this way the equipment is given the protection characteristics of class 0.

Class 0I equipment is extremely rare and only encountered in specialised electrical installations, such as test laboratories. This type of equipment would not be suitable for general use and therefore would constitute a failure during any visual inspection.

CLASS I

A vast majority of early electrical appliances were constructed from metal. Typical examples include simple kettles, irons, toasters etc., so it was a logical progression that to improve the safety of these appliances their exposed metal parts would be connected to the earth in the fixed installation. The first official standard for two-pin and earth plugs and sockets arrived as early as 1928, but the wider standardisation of this type of earthed equipment didn't really take off until the introduction of our current BS 1363 plug and socket in 1947. The introduction of a plug top fuse as part of this design also added an additional level of overcurrent protection. Along with the ring final circuit, the BS 1363 plug and socket arrangement was introduced to aid the boom in post-war construction, spurring the demand for electrical appliances.

So the combination of metal construction techniques and improvements in the design of electrical installations, along with the introduction of a standard plug and socket arrangement, created the type of equipment we now know as class I.

Class I equipment has basic protection provided by its basic insulation, and fault protection is provided by the connection of metal parts to the earth in the fixed installation wiring. For many years class I equipment has formed the vast majority of appliances used in the UK and it is only comparatively recently that developments in materials technology have led to the wider use of double insulated, class II equipment.

Class I equipment, like most fixed electrical installations in the UK, employs the protective measure Automatic Disconnection of Supply or ADS. Under normal conditions the user is provided with basic protection in the form of basic insulation, barriers or enclosures. When a fault occurs, i.e. the basic protection fails and live parts come into contact with conductive metal parts of the appliance, a fault current flows to earth, increasing the current through the protective device causing it to operate and automatically disconnect the supply (ADS) thus rendering the appliance safe.

For an ADS system to work, it is essential that all exposed conductive parts of the appliance have a good, low-resistance connection with earth. The greater the resistance of the earth path, the lower the fault current will be and therefore the longer the protective device will take to operate. So the necessity of a low-resistance earth connection is the Achilles heel of all class I systems.

In addition to typical faults, such as bad earth connections inside the plug and appliance, earth resistance can also be affected by cable length and conductor cross-sectional area, making an earth continuity test an essential part of PAT testing class I appliances. When testing class I appliances, you should also be aware that it is very common to find intentionally un-earthed metal parts and metal parts that are only earthed for functional or screening purposes. Such parts do not form part of the appliance's fault protection and therefore do not require a low-resistance connection to the earth in the fixed installation, as is the case for earthed metal parts. This can be very misleading and it often leads to appliances being recorded as a fail, when in fact no fault is present.

A typical example of an un-earthed metal part would be the guard around the blades of an office fan, this guard provides mechanical protection to the fan blades but is rarely connected to earth; the fan motor, however, is an earthed metal part and therefore requires a low-resistance connection to earth. The identification of earthed metal parts is the most difficult aspect of testing class I appliances and in many cases only those with suitable experience will be able to make the judgement. Where doubt exists advice should be sought from the equipment manufacturer.

How can we identify class I equipment?

Where it is possible to remove the plug top, we will observe three wires connected: line, neutral and earth. The presence of the earth wire is a strong indication that the equipment is class I. The plug of a class I appliance should always have a metal earth pin; this can aid identification. It is often also possible to read the information printed on the side of the mains flex, where the size and number of cores is displayed: a three-core cable generally signifies class I equipment. A successful earth continuity test gives us final confirmation that the appliance is actually class I.

Great care must always be taken when identifying the class of equipment and you must always consider the possibility that the equipment may have been subjected to poor quality DIY repairs in the past, where for example the plug or flex may have been changed and replaced with non-original parts. It has been known for two-core flex to be used on class I equipment and earth wires to be cut out inside class I plugs. Again, where doubt exists the advice of the manufacturer should be sought.

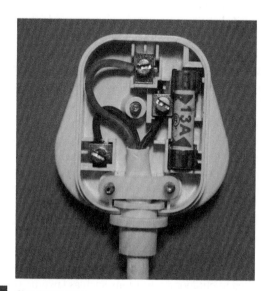

FIGURE 7.9 Class I plug wiring

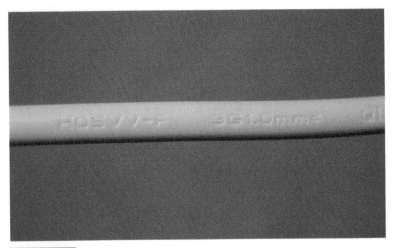

FIGURE 7.10 3G1.0 mm² means the cable has three 1.0 mm² cores

CLASS II

As materials technology evolved, we were able to make insulation in a greater variety of shapes, and as the temperature and UV stability of plastics improved, we were no longer limited to making appliances from metal. This change away from the traditional construction methods caused us to think again about electric shock protection, moving away from the 'earthing' approach and instead placing our reliance on the use of more insulation.

Class II equipment has no connection to the earth in the fixed electrical installation wiring. Instead, we rely on the protective measure of double or reinforced insulation. In class II equipment, basic protection is provided by basic insulation and fault protection is provided by supplementary insulation; however, further developments have also allowed basic and fault protection to be provided by only one layer of reinforced insulation. As class II equipment does not require a connection to earth, we are not reliant on this connection for safety and therefore factors such as lead length no longer pose a potential safety issue, as with class I equipment. This makes class II equipment particularly suited to outdoor applications such as lawnmowers and hedge cutters.

FIGURE 7.11   Class II plug wiring

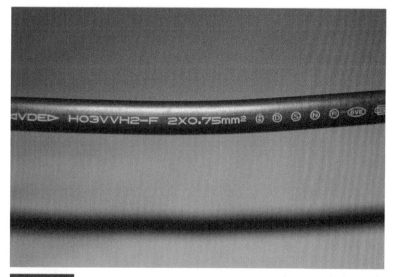

FIGURE 7.12 2×0.75 mm^2 means the cable has two 0.75 mm^2 cores

Class II equipment has a wide variety of construction types and while the traditional view is of a piece of equipment with an entirely plastic enclosure and no exposed metal parts, this is often far from the truth. Many class II appliances such as power tools will have exposed metal parts such as drill chucks, guards, handles and screws, which are separated from live parts by double or reinforced insulation.

We even see items such as DVD players that have a completely metal enclosure; however, inside such equipment live parts are still protected by double or reinforced insulation.

Often, those new to PAT testing will incorrectly identify this type of equipment as class I due to the presence of a metal enclosure.

The identification of class II equipment is usually by the presence of a double insulation construction mark, which is a square inside a square symbol and signifies the two layers of insulation. The *IET Code of Practice* advises that where this mark is not present the equipment must be treated as class I. Other additional methods to aid in the identification of class II equipment include: the absence of a protective conductor inside the plug, a two-core flexible cable and a plastic earth pin.

CLASS III

Where increased levels of protection against electric shock are required, for example in bathrooms or outdoors, it is often necessary to use

FIGURE 7.13 Rating plate with class II construction mark

class III equipment. Class III equipment is specifically designed with safety in mind and offers levels of protection above and beyond that of class I or class II. Class III equipment is supplied from a Separated Extra Low Voltage (SELV) source. The term SELV will be familiar to many electricians as it is widely referred to in the IET wiring regulations.

The term 'separated' refers to the fact that the wiring in a SELV system is electrically separated from earth, making it impossible for the user to receive an electric shock by only touching one conductor or live part while in contact with earth. A shock is, however, still possible if the user simultaneously touches both live conductors. In addition to electrical separation, ELV systems operate at voltages not exceeding 50 V AC or 120 V DC either between conductors or to earth. In the case of SELV the operating voltage is limited for safety reasons and improves protection in the event of electric shock.

Class III equipment must be supplied from a SELV source, usually in the form of a safety isolating transformer complying with *BS EN 61558-2-6* or *BS EN 60742*, where at least double or reinforced insulation is used to separate the primary and secondary windings, preventing secondary winding voltage rising above the ELV voltage limits in the event of a fault. *BS 7671* also lists other SELV sources, such as batteries, generators and electronic circuits provided that they meet the voltage limits and comply with appropriate standards.

Class III equipment is still very rare and limited to safety critical applications. Generally, identification is made by the presence of the class III symbol, which is a roman numeral III inside a diamond. Where an isolating transformer is used to supply class III equipment, confirmation of the correct British Standard, operating voltage and the presence of the safety isolating transformer symbol all must be confirmed. Also, class III equipment will not have an earth connection present.

Typical examples of class III equipment include: modern outdoor Christmas lights and underwater lighting in swimming pools. Many people incorrectly think that all equipment powered via an external transformer, for example mobile phone chargers, are class III. This is usually not the case, unless the supply is via a safety isolating transformer complying with appropriate standards or another SELV source.

Assessment criteria 3.3

Identify relevant construction and identification

There are three equipment construction marks referred to in the *IET Code of Practice* on pages 78 and 79. Typical questions will show you a construction mark and ask you to identify its meaning or describe a piece of equipment and ask you to pick the correct construction mark.

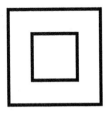

FIGURE 7.14 Class II equipment construction mark

FIGURE 7.15 Class III equipment construction mark

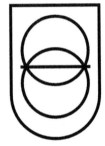

FIGURE 7.16 Safety isolating transformer identification mark

Assessment criteria 3.4

State how electric shock occurs

Electric shock occurs when current passes through the body of a person or livestock. This can occur by contact between live parts or between the live part and earth. The generally accepted level at which we start to notice current passing through our body, the perception level, is about 1 mA and a shock current of just 50 mA is usually fatal.

There are many factors that affect the severity of electric shocks, such as: current level, shock duration and the path taken by the current through the body. Most protective measures aim in the first instance to stop current passing through the body, thus providing basic protection against electric shock. Where the classes of equipment differ is the application of fault protection, where each class uses a different method of protecting us, should the basic protection fail. Class I equipment uses a connection to earth to provide automatic disconnection in the case of a fault; class II equipment relies on supplementary or reinforced insulation; and class III equipment uses electrical separation and a non-hazardous ELV supply voltage.

Assessment criteria 3.5

State methods used to reduce the risk of electric shock from equipment

Whatever system is employed, electric shock can occur in one of two ways, direct contact and indirect contact. Pre-2008 under the 16th edition of the wiring regulations, we commonly referred to the terms 'direct contact' and 'indirect contact'. Following the introduction of the 17th edition in 2008 these terms were replaced by the concepts of basic protection and fault protection. Much confusion has existed since 2008, however, as many people incorrectly translate the old and new terms.

An electric shock by 'direct contact' occurs when a person comes into contact with a part of the equipment that is intentionally live. Such live parts could include: the heating elements inside a toaster, the live conductors inside the appliance flex or the live terminals inside the mains plug. So what protects us against 'direct contact'? To protect us against 'direct contact' we use 'basic protection' measures. For

FIGURE 7.17 Direct contact with the live toaster element

FIGURE 7.18 Indirect contact with exposed conductive parts made live by a fault

portable appliances these usually consist of basic insulation, barriers or enclosures. Therefore we can say that basic protection measures protect us from the parts of equipment that are intended to be live and therefore prevent direct contact.

An electric shock by 'indirect contact' occurs when a person comes into contact with a part of equipment made live by a fault, i.e. a part that would not normally be live during normal use. For example, the metal body of a lamp made live by a fault. So what protects us against 'indirect contact'? 'Fault protection' measures protect us against indirect contact and vary depending on the class of equipment; as mentioned earlier, these could include protective earthing, supplementary or reinforced insulation, electrical separation and the use of non-hazardous supply voltages.

So you can see the relationship between the old and new terms: basic protection protects us against direct contact and fault protection protects us against indirect contact. This relationship is often explored during exam questions, when candidates will be asked to pick an example of the failure of basic protection or the failure of fault protection from a list for given scenarios. To answer this type of question, candidates should identify whether the described scenario is direct contact or indirect contact, i.e. is the part described in the scenario intended to be live or only made live due to a fault?

Assessment criteria 3.6

State how equipment construction protects against electric shock

Below is a table illustrating the four main classes of equipment construction along with their protective measures and main methods of basic and fault protection.

Table 7.1 Protective measures and methods of basic and fault protection in the four main classes of equipment construction

	Class 0	Class I	Class II		Class III
Protective measure	Insulation only	ADS	Double or reinforced insulation		SELV
Basic protection	Basic insulation	Basic insulation	Basic Insulation	Reinforced insulation	Basic insulation
Fault protection	None	Protective earthing	Supplementary insulation		SELV

Assessment criteria 3.7

Explain the effects of conductor resistance

As I mentioned earlier in this book, the resistance of a conductor can be affected by four factors: the conductor material, conductor cross-sectional area, conductor length and the operating temperature. When it comes to our application, portable appliance testing, the main factor that affects conductor resistance is the length of the appliance's supply cable. Many appliances, such as vacuum cleaners, have long cables and it is also very common to see the use of long extension leads, often up to 50 m in length. The conductor resistance increases with length, so a 50 m lead will have 50 times the resistance of a 1 m lead.

Increased conductor resistance can affect appliances in one of two ways: first, class I equipment relies on a low-resistance connection between its earthed metal parts and the earth connection in the fixed installation. If the resistance of the earth conductor increases, this has a limiting effect on the potential fault current that will flow under fault conditions. The level of fault current has a direct effect on the speed at which the protective device will operate. Therefore we can say for class I equipment that the longer the lead, the slower it will take to disconnect in the event of a fault. It is logical to say that once the lead length is increased above a certain point, disconnection within a safe time period will no longer be possible and the fault protection of the appliance will be rendered ineffectual. So, low conductor resistance is critical to achieving fault protection in class I equipment. Where long leads are unavoidable, we often use residual current devices (RCDs) to provide fault protection, as these devices operate in a different way from overcurrent devices, such as fuses or circuit breakers, and will provide protection even when conductor resistances are very high.

The second effect of high conductor resistance is increased voltage drop. As cable resistance increases, there is an increase in potential difference between one end of the cable and the other. When current flows through the cable we see this effect as a drop in the measured voltage between live conductors at the end of the cable; this is referred to as the voltage drop. So even though we may have a measured voltage of 230 V AC at the socket outlet, we will observe a reduced voltage at the appliance. In most cases voltage drop will be negligible and the user will not notice any change in the performance of the appliance; however, if conductor resistance is high, due to for example a long lead, and a high current is flowing in the conductors, voltage drop will be high and the effects can be pronounced. The effects of voltage drop are typically seen as a dimming of lamps, cooling of heating elements and lack of power for motor driven equipment, for example a kettle may take a long time to reach boiling point. To avoid voltage drop problems, cable lengths should be restricted and the cross-sectional areas of cables can be increased.

Assessment criteria 3.8

Identify situations that require the use of RCDs

Residual current devices or RCDs are protective devices that operate on a different principle from the usual overcurrent devices, fuses and circuit breakers used to protect most circuits. An overcurrent device breaks the circuit when the circuit current goes over a set amount for a given time; RCDs, however, operate in a different way.

In a normal, healthy circuit, the current flowing in the line conductor will be equal to the current flowing in the neutral conductor, because in a closed circuit the current flowing in must equal the current flowing out. However, under fault conditions some current may flow to earth. In this case we would see a difference between the line and neutral currents. This difference is referred to as the residual current. A residual current device uses this principle to protect the circuit. The residual current device operates when the residual current rises above a preselected value and disconnects the supply to the circuit. By operating in this way a residual current device can react to very small fault currents, usually in the order of a few milliamps, therefore offering excellent protection against electric shock. However, due to the way that an RCD works, it will not detect overcurrents and for this reason it is often paired with an overcurrent device such as a fuse or circuit breaker.

Residual current devices used to provide additional protection against electric shock must be rated 30 mA or less. Typically RCDs used with portable appliances will have a rated residual operating current ($I\Delta n$) of 30 mA. This means they must operate when the difference between the line and neutral currents is 30 mA or higher, in practice most RCDs will have a tripping point at somewhere between 15 and 30 mA.

FIGURE 7.19   Mains RCD

Residual current devices are usually used to provide additional protection for circuits or equipment with increased levels of risk, which require protection in addition to the usual basic and fault protection methods. The 17th edition of the wiring regulations (BS 7671) explains that additional protection, by the use of an RCD rated 30 mA or less, is used to provide protection in the event of the failure of the basic and/or fault protection or carelessness by users. In some cases, such as the long extension leads mentioned earlier, the RCD is also used to provide fault protection.

Residual current devices take many forms: they can be mounted in the distribution board or consumer unit; be incorporated into socket outlets or fused connection units; or simply be plug-in devices designed for use with portable appliances. Each residual current device will provide protection to wiring and equipment downstream of that device. In practice it does not matter whether the appliance is protected by an RCD in a consumer unit or one plugged into a socket outlet, the protection afforded should be the same. Where the user of an appliance cannot be certain that a socket outlet is RCD protected they may prefer to use a plug-in device to ensure additional protection of the appliance.

There are many types of installation and appliance for which RCD protection is recommended, most of these requirements are detailed in the 17th edition wiring regulations (BS 7671) and can be quite technical in nature, so if any doubt exists the advice of a suitably qualified electrician should be sought.

Typical examples include:

- Equipment used outdoors.
- Equipment in bathrooms or locations where water is present, such as swimming pools.
- Equipment with long cables and long extension leads.
- Cables concealed in walls.
- Equipment with an increased risk of electric shock.

Where used, residual current devices will greatly contribute to improving electric shock protection. They should, however, be functionally checked before use by pressing the built-in test button. Further testing must also be carried out during the in-service inspection and testing process, this will confirm that the device operates within the required time limits.

CHAPTER 8

Learning Outcome 4

Learning outcome 4

Understand the procedures for the in-service inspection and testing of electrical equipment

The title of this learning outcome is a bit misleading, as the assessment criteria are mainly concerned with understanding frequencies at which inspection and testing should take place. We will also look at the visual inspection process undertaken by the user and as part of the formal visual inspection.

Assessment criteria 4.1

Identify the types of in-service inspection and testing

This topic was covered in more detail in Chapter 2, earlier in this book. The criteria relates to the types of in-service inspection and testing, which essentially break down into three areas:

1. *User check* – a basic visual inspection carried out by the user, which is not recorded unless a fault is found.
2. *Formal visual inspection* – a more detailed visual inspection carried out by a competent person, the results of which are always recorded.
3. *Combined inspection and test* – again carried out by a competent person, this type of inspection combines the formal visual above with a series of electrical tests; the results of combined inspection and test are always recorded.

The above are carried out at different layered intervals, such that user checks have the greatest frequency, followed by formal visual inspections, and finally combined inspection and test has the lowest frequency; the aim being to provide a comprehensive system of in-service inspection and testing, which detects the most common faults quickly, while reducing the burden placed on the organisation by frequent combined inspection and testing.

Assessment criteria 4.2

State the factors which determine the initial frequency of inspecting and testing

This criteria was also addressed earlier in the book, during Chapter 2, where we discussed the nature of risk and looked at the factors that need to be taken into account. The *IET Code of Practice* covers this topic in chapter 7.3 on p. 50.

When assessing the frequency of inspection and testing it is important to consider the legal requirement to maintain equipment to prevent danger. Regulation 4(2) of the *Electricity at Work Regulations 1989* uses the phrases 'as may be necessary to prevent danger' and also 'so far as is reasonably practicable'. So legally we are asked to make a judgement about frequency of inspection and testing. The interval between inspection and testing must be short enough to prevent danger but also reasonable and practical. This is a difficult determination with no specific 'right answer'. As mentioned earlier, the frequency at which each of the three types of in-service inspection and testing are carried out must be based on an assessment of risk.

Each type of in-service inspection and testing is a control measure utilised by the organisation to reduce risk to acceptable levels. When performing risk assessments, we must consider all relevant factors, and the more factors we use, the more accurate the risk assessment will be. The *IET Code of Practice* lists seven factors and these should be your focus during your exam preparation, with regards to this assessment criteria.

The factors listed are:

- **The environment** – e.g. where is equipment to be used? Are there any special considerations, such as the presence of water?
- **The users** – e.g. who will use the equipment, what levels of skill or training will they have? How will this impact the risk?
- **The equipment construction** – e.g. class I equipment may require more frequent testing due to its reliance on a connection with earth.
- **The equipment type** – e.g. hand-held equipment poses a greater risk to the user, therefore it may require more frequent testing.
- **The frequency of use** – e.g. items that are used frequently will wear out sooner; an item used only once a year will not require frequent testing.
- **The type of installation methods** – e.g. long, trailing mains cables may be subject to damage, while cables installed in trunking or behind equipment are unlikely to receive the same amount of damage.
- **Previous records** – e.g. by studying previous records we can establish patterns or trends that may be able to help us establish suitable frequencies.

Assessment criteria 4.3

Identify the initial frequencies of inspection and testing of equipment

The *IET Code of Practice* is careful to use the phrase 'initial frequencies' when referring to the frequencies listed in table 7.1 on p. 52, and it is made clear that these frequencies should not be considered as an absolute legal requirement. Table 7.1 is a very important part of the *IET Code of Practice* and will be heavily referenced during your exam. The frequencies listed are included for guidance purposes and can be used as a benchmark to calibrate your assessment of risk against those frequencies suggested by the IET. Ultimately, your frequencies must be established by your own assessment of risk and, when push comes to shove, you should be confident that you could defend your choices in court. Therefore it is

FIGURE 8.1   Frequencies must be reviewed at regular intervals

always advisable to have written risk assessments that detail the process
you used to establish the level of risk and your control measures.

Table 7.1 in the *IET Code of Practice* is divided up into seven columns
covering: the equipment environment, the type of equipment, suggested
frequency of user checks, suggested frequency of formal visual inspections
and combined inspection and tests for both class I and class II equipment.
It is essential for your exam that you practise using table 7.1 to establish
the initial frequencies for different types of equipment. Practise by setting
yourself different examples and selecting frequencies from the table.
This type of question should be very easy, but often people make silly
mistakes, so always double-check your answers to make sure.

To make things a little bit more difficult, examiners often ask questions
that relate to the notes that accompany table 7.1, listed on p. 53 of
the *IET Code of Practice*. Some typical examples include note 2, which
requires 230 V portable and hand-held equipment on a construction
site to be RCD protected and inspected and tested 'more frequently'
than 110 V equipment. Note 3 mentions 'children's rides' and states
that a daily user check may be necessary. Note 4 reminds us that in
certain environments, for example equipment used by the public, we
cannot expect the user to carry out a user check, therefore checks would
be carried out by a nominated supervisor, teacher or member of staff.
Finally, note 5 clarifies that equipment in hotel rooms is classified as

equipment used by the public (row 3 in table 7.1), while in row 5 in table 7.1 'hotels' only refers to equipment in hotels, used by hotel staff.

As you can see, the above notes can lead to confusion during exam questions and it is important that you thoroughly familiarise yourself with these notes before attempting this type of exam question.

It is also very important to remember that frequencies of inspection and testing must be kept under review and any changes in the level of risk are quickly identified and the necessary changes made to your programme of inspection and testing.

Assessment criteria 4.4

Specify what needs to be considered when carrying out a formal visual inspection

In order to show due diligence, formal visual inspections must be recorded. The *IET Code of Practice* contains sample forms in appendix V. Form V.2 (equipment formal visual and combined inspection test record) is an example of the type of form used to record formal visual

FIGURE 8.2 Formal visual inspections must always be recorded even if the item fails, as this demonstrates due diligence

inspections. During your course practical assessment you will be asked to complete this type of form, so I would recommend familiarising yourself with its layout and practising completing this form for several practice appliances prior to undertaking the practical assessment itself. Boxes on form V.2 are labelled with notes 1 to 19 detailed on the reverse of the form; these notes explain what information must be entered in each box and this is a great help when learning how to fill in this form. When performing a formal visual inspection only (i.e. without tests), all boxes are completed with the exception of the test result boxes (15 i to v). Care must also be taken when completing this form to adhere to the instructions at the bottom of the form, which remind us that ✓ indicates pass, ✗ indicates fail, N/A not applicable and N/C not checked. Also, Y and N are used for yes and no. Candidates who do not use these symbols to complete the form, instead opting for similar terms such as OK or YES, may risk being marked down or even fail the practical assessment. No boxes in the column should be left empty.

When performing a formal visual inspection, safety is very important because the condition of the equipment is as yet unknown, so may pose a potential danger to the inspector. Where available, the inspector should review any previous test records, which will allow them to familiarise themselves with the appliance and its previous condition, so that any deterioration or re-occurring faults may be identified.

The inspector may also find it useful to review any manufacturer's instructions, which will give information on the correct use of the equipment, environments for which it is suitable and other useful technical information such as construction class. Where the user of the equipment is known to the inspector, the inspector should consult the user, who will be aware of equipment defects, such as intermittent faults, overheating and other equipment problems. The inspector must consider this information before proceeding with visual inspection, where judged safe to do so.

Safe isolation of equipment prior to conducting the formal visual inspection is an essential safety precaution. The inspector should determine that the item of equipment can be safely disconnected from its supply and, where permission is required, this must be obtained before equipment is isolated. The inspector must also be aware that some types of equipment may be powered by backup sources such as uninterruptible power supplies (UPS) or backup generators.

FIGURE 8.3 Uninterruptable power supply

Before inspection, the equipment must also be disconnected from any backup sources. Again, permission must be obtained for this type of isolation. Safe isolation may be simple to achieve if appliances are connected to their supply via a plug and socket arrangement; however, items directly connected to the mains electrical supply, i.e. fixed equipment wired to a fused connection unit, will be more difficult to isolate safely, and the increased level of competence required may mean that this is a job for a qualified electrician.

It is also important to isolate equipment from any communications connections, such as telephone lines, networks, aerials etc. The Inspector will require permission before disconnecting these connections and may also need the help of a suitably qualified user, such as a member of IT support staff.

The *IET Code of Practice* also mentions the potential dangers posed when disconnecting fibre-optic connections. This type of disconnection should only be undertaken by suitably trained staff familiar with the dangers and precautions required. If any equipment cannot be safely isolated, inspection and testing must not be carried out and this should be brought to the attention of the responsible person.

Once the inspector has determined the equipment has been safely isolated from its supply, the formal visual inspection can start. The *IET Code of Practice* contains a list of typical items to be inspected in table 13.1 on p. 82. While these items are intended to form the user check, they are essentially the same items inspected during the formal visual inspection. Until you build up suitable experience of performing formal visual inspections, I would suggest using table 13.1 as a checklist to work through step-by-step while inspecting each item. After a period of time inspecting, these items will become second nature.

When performing a visual inspection you should use a methodical approach, gradually working your way through the various parts of the plug, cable and appliance confirming the serviceability of each item step-by-step. Your aim should be mentally ticking off each item as okay, not focusing solely on trying to spot faults. The vast majority of items that you check each day will have no faults, so an inspector who concentrates on just looking for faults will often become complacent and miss faults when they do arrive. A good inspector will check every item in the same way using their mental tick list, ensuring all faults will be identified naturally as part of this process. In addition to the items in table 13.1, formal visual inspection will also involve additional items, such as removal of the plug top and inspection of the internal connections. The *IET Code of Practice* gives further guidance in appendix VII, starting on p. 135.

When inspecting the internal wiring of the plug, it is important to adopt the same methodical approach. Most faults inside the plug result from poor-quality DIY termination methods performed by users. We must also be on our guard for poor quality repairs, such as replacements of the fuses with incorrect rating, fuses wrapped in tinfoil and even, in some cases, bolts or nails used to replace the fuse. Those new to visual inspection should review appendix VII in detail as it lists the typical defects you may come across while conducting this type of inspection.

The plug top fuse plays an important role in ensuring the safety of the appliance, therefore it is essential when conducting a formal visual inspection that we confirm that the fuse is of the correct type and rating. The fuse, sometimes referred to as a cartridge fuse because of

FIGURE 8.4 BS 1362 fuse with BSI and ASTA marks

its construction, should be removed from the plug to allow its end
caps and the terminals within the plug to be inspected, making sure
there is good tight fit between the plug terminals and the fuse and
that no discolouration or overheating is apparent. The fuse should
be checked to ensure it is marked with the British Standard BS
1362. Also present should be an ASTA mark and a British Standard kite
mark.

In recent years there has been a large influx of counterfeit electrical
equipment, and counterfeit fuses, plugs and leads are sadly very
common. While the presence of the BS number and appropriate
marks are not proof that fuses are genuine, any fuses not bearing this
information must be treated as suspect, also casting doubt on the
genuine nature of the plug, cable and appliance.

The *IET Code of Practice* (15.13, p. 101) includes guidance on the rating
of plug top fuses to BS 1362, which is detailed below:

- Appliances up to 700 W should be fitted with a 3A fuse (red).
- Appliances over 700 W should be fitted with a 13A fuse (brown).

The above guidance is intended as a rough rule of thumb; however,
manufacturers' recommendations should always be followed and
fuses should not be changed from their original manufacturer's
rating. Manufacturers may decide to increase fuse sizes for operational
reasons such as start-up or inrush currents. A typical example of
this is that 650 W drills are commonly supplied with a 13 A fuse,
the manufacturer having determined that a 3 A fuse would not be
suitable due to the increased currents present during start-up. Where a
plug is of the moulded on type, i.e. the plug top cannot be removed, the
correct fuse rating, often 5 A for IT equipment, will be marked on the
plug top. The inspector should always confirm that the correct fuse is
fitted.

FIGURE 8.5 3A and 13A fuses

FIGURE 8.6 Moulded plug marked to indicate that a 3A fuse should be fitted

The rating of fuses is often used as a subject for exam questions. Candidates are asked to select the correct fuse size for the appliance described in a given scenario. Candidates are also often asked, what does the plug top fuse protect? The *IET Code of Practice* states that the plug top fuse protects the flex against faults.

The *IET Code of Practice* also gives guidance on the length/CSA of appliance leads (15.12) and extension leads (15.10), which should be considered during the visual inspection.

Table 15.6 on p. 101 of the *IET Code of Practice* states minimum cross-sectional areas for appliance flexes protected by 3 A or 13 A fuses. We are reminded that with flexes protected by these fuse sizes and meeting the minimum cross-sectional areas, 0.5 mm² and 1.25 mm² respectively, no limit is placed on flex length. However, as mentioned earlier in this book, voltage drop may require flex lengths to be restricted.

Table 15.4 on p. 98 of the *IET Code of Practice* gives the recommended maximum lengths for site-manufactured extension leads, based on lead cross-sectional area:

- 1.25 mm² cable – maximum length 12 m
- 1.5 mm² cable – maximum length 15 m
- 2.5 mm² cable – maximum length 25 m

The above maximum lengths apply to extension leads only and are not applicable to appliance flexes. Where cable length exceeds the above, RCD protection should be provided due to the increased protective conductor resistance. We must also be aware that 2.5 mm² cable is too large for a standard BS 1363 plug; however, this type of cable may be used with a BS EN 60309 industrial type plug.

In addition to the requirements to visually inspect the appliance, the inspector must also consider the wider safety implications of the environment in which the appliance will be used and the tasks it will be required to carry out.

Electrical equipment must be suitably rated for safe operation in the chosen environment. The inspector must consider, where known, any external influences applicable to the appliance's operating environment. Typical examples include: impact, adverse weather conditions, high or low temperatures, presence of water or the proximity of flammable or explosive substances. Once applicable external influences have been identified, the inspector must determine that the appliance is suitable for safe operation by inspecting the appliance and consulting manufacturers' information.

Many appliances will have an IP rating, which details the level of ingress protection (IP) offered by the appliance. IP codes usually consist of the letters 'IP' followed by two numbers, for example 'IP44'. The first number relates to the protection provided against solid foreign objects, ranging from 0 (no protection) to 6 (dust tight). The second number relates to the protection provided against the ingress of water, ranging from 0 (no protection) to 8 (protection against continuous immersion in water). More details on IP codes are given in the definition section of the *IET Code of Practice*, tables 2.1 and 2.2, p. 28. Exam questions often refer to IP codes, where the candidate will be given a code and asked to pick from four possible meanings, or asked to pick the correct code from an appliance description. Practice of this type of question is recommended during your revision.

Whenever possible, the inspector should consider the suitability of the appliance for the task it will be used to carry out. If an appliance is not suitably rated for a particular task it may experience high levels of wear or even catastrophic failure posing a significant danger to the user. For example, a DIY drill may state on its rating plate that it is only suitable for use with drill bits of 10 mm or less in diameter. If this drill were used with a 25 mm diameter hole saw it may become overloaded

FIGURE 8.7 Appliance with IP55 rating

FIGURE 8.8   Drill with maximum drill size of 10 mm

and overheat, causing the insulation to fail. If the inspector feels that the item is not suitable, this must be brought to the attention of the responsible person.

The safety of many electrical appliances depends heavily on the safety of the fixed electrical installation. This is particularly the case for class I equipment, where electric shock protection relies on earthing provided by the fixed installation. In addition to the in-service inspection and testing of appliances and equipment, the *Electricity at Work Regulations 1989* (regulation 4(2)) also require that the fixed electrical installation is maintained in a safe condition. Each electrical installation will have a label at its origin stating the date of the last periodic inspection and the recommended next inspection date. The inspector should determine whether the electrical installation has a current periodic inspection. Where the installation does not have a current periodic inspection, this must be brought to the attention of the responsible person and great care must be taken by the inspector during inspection and testing activities to ensure that any defects in the electrical installation do not pose a danger to safety. During the formal visual inspection, the inspector must also, where possible, visually inspect any socket outlets to which the appliance may be connected for signs of damage, overheating or overloading. Again, any defects must be reported to the responsible person without delay.

Part of the formal visual inspection that is often overlooked is the need to confirm the effectiveness of the means of isolation and switching for each appliance. Paragraph 14.3 on p. 86 of the *IET Code of Practice* details the switching requirements. Essentially the inspector should determine that, where necessary, the appliance has effective means of switching for normal functional use, to carry out maintenance and in the event of an emergency. In addition to confirming that the switching devices operate correctly, we must also ensure that these devices are readily accessible and clearly identified. Typical examples of problems that you may encounter include: inaccessible floor sockets covered by furniture; equipment such as photocopiers located in front of socket outlets, blocking access; and equipment located high on walls or ceilings where users are unable to reach the means of switching and may be tempted to stand on chairs or tables, making operation of the equipment unsafe.

CHAPTER 9

Learning Outcome 5

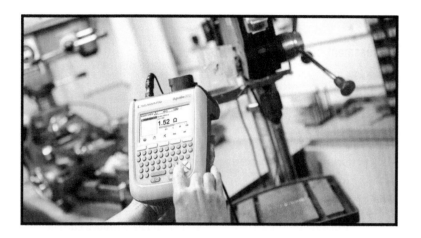

Learning outcome 5

Understand how to carry out combined inspection and testing

This very important learning outcome allows us to get to grips with the electrical tests carried out as part of portable appliance testing. Electrical testing is used to supplement visual inspection – by carrying out electrical tests we can confirm the safety of items that it is not possible to verify by visual inspection alone. For example, it is impossible to see the protective conductor inside a length of cable, so instead we measure its electrical resistance and use this information to evaluate its condition.

When carrying out combined inspection testing on electrical equipment or appliances it is important to remember that a large majority of faults will only be discovered by thorough visual examination. Electrical testing is important, but it only gives us a small amount of information, which must be correctly interpreted if it is to add to our overall understanding of the safety status of the appliance. For electrical test results to be meaningful, testing must have been carried out in the correct

Get Qualified: Portable Appliance Testing. 978 1 138 18955 3 © Kevin Smith, 2017. Published by Taylor & Francis.

way. Anyone wishing to carry out electrical testing must familiarise themselves with the range of tests available, understand how to apply each test correctly and interpret the results obtained. The key to good PAT testing is knowing the right tests to apply to each item of equipment to get accurate, meaningful results.

Assessment criteria 5.1

Identify tests that are suitable for the different types and classifications of in-service equipment

The *IET Code of Practice* covers combined inspection and testing in chapter 15, which starts on p. 89; paragraph 15.3 lists the tests required when performing in-service testing of electrical equipment.

The basic tests required by the *IET Code of Practice* are:

- An earth continuity test (class I equipment only).
- An insulation resistance test, or protective conductor/touch current tests or alternative/substitute leakage tests.

The above tests should, if passed, then be followed by a 'functional check', during which the item is connected to its supply, switched on and run for a sufficient period of time in all modes of operation to confirm that it performs safely and poses no danger to the user.

When carrying out in-service inspection and testing of any item of electrical equipment, we must always consider the item to be potentially dangerous until we have proved otherwise by a rigorous process of visual inspection, electrical testing and functional checks. It is for this reason that the order in which we carry out these activities is critical and we only progress to the next step when the equipment passes each stage.

1. **Visual inspection** – With the item safely isolated from its supply we perform a thorough visual examination.
2. **Dead testing** – We perform electrical tests on the equipment using test voltages generated by the PAT tester; these tests may include:
   - Earth continuity test (class I only)
   - Insulation resistance test
   - Substitute/alternative leakage test
   - Polarity test

3. **Live testing** – We perform mains powered electrical tests on the equipment using the PAT tester; these tests may include:
   - Protective conductor current test (class I)
   - Touch current test (class II)
   - RCD test
   - Load test
4. **Functional check** – A final check to ensure that the equipment operates safely and in accordance with the manufacturer's instructions.

By following the sequence illustrated above we gradually prove the safety of the equipment, step-by-step. Failure to follow this procedure, for example by performing live tests at 230 V before a thorough visual examination has been carried out, could pose a serious danger to the tester and others in the test area.

IN-SERVICE INSPECTION AND TESTING

Below is a table detailing the typical sequence of inspections and tests applied to appliances and leads:

Table 9.1 Typical sequence of inspections and tests

	Appliance		Lead	
	Class I	Class II	Class I (3 – Core)	Class II (2 – Core)
Visual inspection	✓	✓	✓	✓
Earth continuity	✓	N/A	✓	N/A
Insulation resistance	✓	✓	✓	✓
Polarity	N/A	N/A	✓	✓
Substitute/alternative leakage	Optional	Optional	Optional	Optional
Protective conductor current	Optional	N/A	Optional	N/A
Touch current	N/A	Optional	N/A	Optional
Load	Optional	Optional	N/A	N/A
RCD	Where present, an RCD should be functionally checked and tested to confirm it meets required disconnection times.			
Functional check	✓	✓	✓	✓

NOTE: The above table is a basic guide; however, the tester must always make the decision of which tests are applicable for each item. It may not be possible to carry out all tests on every item of equipment. For example, tests may need to be omitted for equipment with no exposed conductive parts.

Assessment criteria 5.2

Identify the range of test instruments that can be used for testing electrical equipment.

This assessment criteria relates to test instruments, a topic covered in chapter 10 of the *IET Code of Practice*, which starts on p. 63. It is clear that if we are hoping to carry out portable appliance testing, beyond a simple visual inspection, we are going to need some form of electrical test equipment. It makes sense to use a dedicated portable appliance tester, i.e. a piece of test equipment specifically designed for the task. It is, however, possible to carry out portable appliance testing with suitable stand-alone test instruments and, in some cases, for example fixed equipment, the use of stand-alone test instruments may be more appropriate.

All equipment used to carry out portable appliance testing must be safe to do so and the test operative must be suitably trained to operate the equipment and interpret the results obtained. The current safety standard for this type of equipment is *BS EN 61010*, and all new equipment should comply with this. Further information on the safety of test leads and probes can be found in *HSE guidance note GS 38*.

In simple terms, a portable appliance tester is a piece of electrical test equipment, usually fitted with a BS 1363 three-pin socket outlet to facilitate the easy connection of appliances. The portable appliance tester will have at least the ability to carry out earth continuity and insulation resistance testing. There is a huge range of portable appliance test equipment available, ranging from simple PAT checkers, which only give a pass/fail indication usually in the form of a red or green light, to comprehensive pieces of test equipment that offer a wide range of tests, even in some cases incorporating a camera allowing faults to be photographed and stored. It is important when deciding to purchase a portable appliance tester that you consider your application carefully: What type of appliances will you be testing? How many appliances? And what type of records do you want to keep?

Portable appliance testers generally fall into two categories: non-downloading and downloading. Non-downloading testers have no facility to allow you to record your test results within the tester. When using this type of tester, results must be recorded on paper or using another device such as a tablet or laptop. This type of tester is better suited to low-volume testing of simple appliances. Downloading testers allow you to capture information about the appliance and the test results, which can then be downloaded to a computer at a later date. If you are looking to test large numbers of more complicated appliances and provide computerised reports to your client, a downloading tester is a must.

As with purchasing any item, you should expect to pay more for portable appliance testers with a wider range of features. Below, I have put together a list of common portable appliance tester features. When aiming to buy a tester you should identify which of these features you need and purchase a tester that meets your specification.

Common portable appliance tester features:

- Simple pass/fail indication.
- Displays measured values (i.e. 0.09 Ω not just PASS).
- Factory set pass/fail limits only.
- Variable pass/fail limits.
- Battery powered operation or mains only.
- Fixed or selectable earth continuity test current.
- Fixed or selectable insulation resistance test voltage 500/250 V.
- Substitute or alternative leakage tests.
- Mains powered leakage tests.
- An IEC socket for lead testing.
- The ability to test 110 V or three phase appliances.
- RCD testing.
- Load testing.
- The ability to store test results.
- User programmable test sequences.
- Barcode scanner for data entry and label printer.
- Built-in camera.
- Database and reporting software available.

When it comes to buying a PAT tester I would always advise taking your time and doing plenty of research. The rule of thumb is to buy the best tester you can afford, as you are unlikely to complain if your tester has too many features; however, buying a cheap tester may slow you down and seriously limit the range of items you can test. Often drawbacks are not apparent until you start using the tester.

FIGURE 9.1   Seaward Apollo 600 with built-in camera

Once you have decided which tester to buy, you can then start shopping around to get the best deal, but never simply buy on price alone. It is often possible to find testers included as part of a kit, bundled with extra accessories, labels, training and software kits, which usually offer good value, as buying these items separately later is usually very expensive. I would always recommend buying from manufacturer approved distributors only, where after sales support is likely to be much better. Also ensure when buying a new tester that it comes with a current certificate of calibration, which will cost you extra if not included.

The *IET Code of Practice* also recommends that test operatives should be trained so that they are familiar with the test instrument used, particularly its limitations and restrictions, so that repeatable results can be achieved without damage to the equipment. Test equipment manufacturers offer a range of training courses based on their products, and this type of training is often invaluable, allowing you to get up-and-running quickly and to get the most out of your portable appliance tester.

PAT TESTING WITH STAND-ALONE INSTRUMENTS

As mentioned earlier, it is perfectly possible to carry out portable appliance testing with stand-alone test instruments. The *IET Code of Practice*, paragraphs 10.3 and 10.4 starting on p. 65, explain that earth continuity testing may be carried out using a **low-resistance ohm meter** conforming to BS EN 61557–4 and insulation resistance testing may be carried out using an **insulation resistance ohm meter** complying with BS EN 61557–2. The use of these instruments for PAT testing is very rare, but as it is possible examiners occasionally use this as the subject for exam questions (e.g. What item of test equipment could be used to carry out an earth continuity test?)

Assessment criteria 5.3

Explain the need for test instruments to be calibrated and in good working order

In order for us to rely on the results generated by a portable appliance tester we must be certain that they are accurate, they must comply with the requirements of the manufacturer and any applicable British or harmonised standards. In order to confirm the above, the portable appliance tester must be calibrated in an approved calibration laboratory, where it will be subjected to a predetermined range of tests in controlled laboratory conditions. On successful completion of the calibration process you will receive a certificate confirming that the portable appliance tester was within the required parameters at the time of test; this certificate is not, however, a guarantee that the portable appliance tester will remain accurate until its next calibration.

To ensure the ongoing accuracy of your portable appliance tester it is important to ensure that the item is used and stored in accordance with the manufacturer's instructions. The *IET Code of Practice*, paragraph 10.5 p. 67, further recommends the portable appliance tester is subject to routine inspection and testing to ensure its ongoing accuracy. Form V.6, in appendix V p. 132, gives an example of an instrument test record, which should be maintained for each test instrument. Formal calibration should be carried out annually or in accordance with manufacturer instructions; however, a test for ongoing accuracy should be performed more frequently, for example weekly or monthly, to ensure the portable appliance tester remains accurate for the period between calibrations. The purpose of the test for ongoing accuracy is to identify any change in the readings between one test and the next; these checks will also allow us to identify any inaccuracy or damage to the instrument caused by day-to-day use.

It is important that prior to using any test equipment we carry out a thorough visual examination of the equipment, any accessories and leads for signs of damage or deterioration that may make the equipment unsafe or unsuitable for use. Equipment found to be in this condition must be removed from service immediately and returned to the manufacturer for repair and recalibration.

Assessment criteria 5.4

Calculate the resistance of the flexible cable conductors and protective conductors

When we carry out a protective conductor continuity test, the portable appliance tester will provide us with a reading of measured conductor resistance in ohms. If this value is higher than expected it may indicate

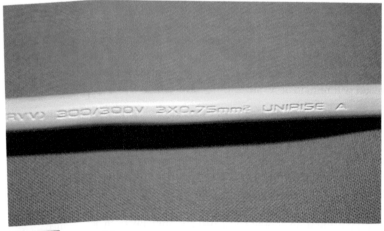

FIGURE 9.2   0.75 mm² three-core cable

poor earth connections in the plug or appliance or damage to the
conductors inside the cable. Before we can make this judgement we need
to calculate what the resistance of the cable should be; this value can
then be compared with the measured value.

Earlier in this book we learned that the resistance of a conductor is
dependent on the conductor's cross-sectional area and its length. The
cross-sectional area of a conductor is usually marked on the outside of
the cable sheath and will be measured in mm², see Figure 9.2.

To establish the length of a conductor this can either be estimated to
the nearest metre or measured with a tape measure. We can then use
data taken from cable manufacturers to calculate the overall conductor
resistance. The *IET Code of Practice* has included table VI.1 on p. 134,
which contains information on the standard cable sizes we are likely to
encounter during portable appliance testing.

Below is a summary of the common values likely to be needed during
day-to-day PAT testing and during your exam:

Table 9.2 Common resistance values likely to be needed during the calculation
of earth continuity pass/fail limits

Conductor CSA	mΩ/m	Ω/m
0.75 mm²	26	0.026
1.0 mm²	19.5	0.0195
1.25 mm²	15.6	0.0156
1.5 mm²	13.3	0.0133
2.5 mm²	8	0.008

To calculate conductor resistance we simply multiply the Ω/m value by
the length in metres. Below are some typical examples:

- *Example 1* – **Calculate the resistance of a 1.25 mm² cable with a
  length of 10 m.**

A 1.25 mm² cable has a resistance of 15.6 mΩ/m, this value can be
converted into Ω/m by dividing it by 1,000.

$$15.6 \text{ mΩ/m} \div 1,000 = 0.0156 \text{ Ω/m}$$

This value is then multiplied by 10 as the cable length is 10 m.

$$0.0156 \text{ Ω/m} \times 10 \text{ m} = 0.156 \text{ Ω}$$

Therefore the resistance of a 1.25 mm² cable with a length of 10 m is calculated to be **0.156 ohms**.

- *Example 2 – A cable of CSA 1.5 mm² and 1 m in length has a resistance of 0.0133 Ω. What would be the resistance if this cable were extended to 5 m?*

If 1 m of 1.5 mm² cable has a resistance of 0.0133 Ω, then 5 m will have 5 times this resistance.

$$0.0133 \ \Omega \times 5 = 0.0665 \ \Omega$$

Therefore the resistance of a 5 m lead with the cross-sectional area of 1.5 mm² is **0.0665 Ω**.

- *Example 3 – A cable of CSA 1.5 mm² and 12 m in length has a resistance of 0.1596 Ω. What would be the resistance if this cable CSA were reduced to 0.75 mm²?*

If the cable CSA is halved the resistance will double.

$$0.1596 \ \Omega \times 2 = 0.3192 \ \Omega$$

Therefore, using the above method, the resistance of a 12 m lead with the cross-sectional area of 0.75 mm² is **0.3192 Ω**. Some calculated values may differ due to rounding errors and the numbers of decimal places used – you must keep this in mind when selecting the correct multiple-choice answer in your exam.

Assessment criteria 5.5

Specify how to carry out earth continuity testing

The *IET Code of Practice* addresses earth continuity testing in paragraph 15.4 starting on p. 91.

The safety of all class I equipment relies on a good, low-resistance, electrical connection between earthed metal parts and the earth connection in the fixed electrical installation. It is therefore essential that we confirm this connection by measuring the resistance between all earthed metal parts and the equipment's earth connection, usually the earth pin on the mains plug or the earth terminal within a fused connection unit or cooker outlet. A low resistance reading signifies a good electrical connection; however, a high resistance reading may indicate a poor connection within the plug or the appliance, or damage to the protective conductor within the cable.

Below is a diagram showing an earth continuity test on a simple class I appliance:

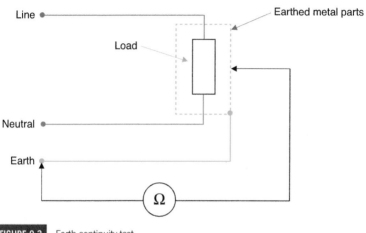

FIGURE 9.3 Earth continuity test

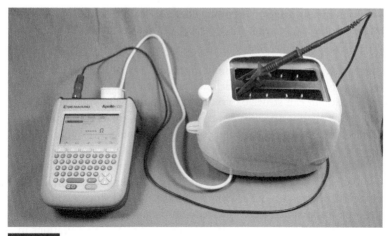

FIGURE 9.4 Earth continuity test connections example

The resistance measurement is made between the earth pin of the plug and earthed metal parts.

The *IET Code of Practice* lists two methods for carrying out an earth continuity test: the 'hard' test and the 'soft' test. In truth these two tests are very similar and the only notable difference is the amount of current used to take the measurement.

The 'hard' test uses a high test current, at least 1.5 times the rating of the fuse up to a maximum of about 26 A. The test current is so high that the duration of the test must be limited to between 5 and 20 seconds to avoid overheating. As the name suggests, the 'hard' test will subject the earth path to a certain amount of stress; this is, however, a double-edged sword. Applying a high current to the earth path allows us to obtain a reliable and stable reading, even through usually problematic contaminants such as lime scale, rust and paint. The 'hard' test also gives us a greater degree of confidence that the earth path will be able to stand up to the conditions that may be present during an actual earth fault.

Unfortunately, applying this high current to sensitive parts of an appliance, such as printed circuit boards or wires with a small cross-sectional area, may cause serious damage to the appliance. Due to this high risk of damage, the 'hard' test should only be carried out when the tester is sure that the earth path can carry such high currents. Typical examples include extension leads and robust appliances with solid metal casings. Where any doubt exists, the 'hard' test should not be carried out; instead, the appliance should be subjected to the 'soft' test.

The 'soft' test is an earth continuity test made with a test current within the range of 20 to 200 mA. This lower test current is unlikely to damage equipment, but can struggle to get a stable reading when the earthed metal parts are not perfectly clean. As mentioned above, contaminants such as lime scale, rust and paint can increase contact resistance making the earth continuity reading artificially high or unstable. To reduce the effects of contact resistance, most modern PAT testers use test currents of 200 mA but it may still be necessary for the tester to clean the earthed metal part before a test measurement is made. To help identify poor earth connections it is recommended that the appliance cable is flexed whilst carrying out an earth continuity measurement. Any variations in the reading during such flexing should be investigated.

Because of the risk of damage posed by the 'hard' test, it is very common nowadays for all testing to be carried out using the 'soft' test method. In fact, many modern battery powered PAT testers don't have a 'hard' test function at all. Some parts of appliances are only earthed for functional or screening reasons and not for protection against electric shock. While these items can be checked with the 'soft' test, they should never be subjected to the 'hard' test, which is likely to result in damage to the appliance.

You should also be aware, when testing class I appliances for earth continuity, that not all metal parts are connected to earth. Un-earthed metal parts are very common and those who are new to portable appliance testing may struggle to distinguish between earthed parts that require testing and un-earthed parts that don't. A typical example of an un-earthed part is the metal guard around the fan blades of an office fan, which is usually not earthed and simply exists to provide mechanical protection from the rotating fan blades, while protection against electric shock is provided by the earth connection to the metal body of the fan motor. Some class I equipment may also have no accessible earthed metal parts, usually due to a comprehensive plastic enclosure limiting access to earthed parts, therefore making earth continuity testing impossible. Where no earthed parts are accessible to the tester, testing cannot be carried out and this should be recorded and brought to the attention of the responsible person.

EARTH CONTINUITY PASS/FAIL LIMITS

Table 15.1 on p. 93 of the *IET Code of Practice* gives the pass/fail limits for earth continuity testing of appliances and leads.

Earth continuity pass/fail limit:

$$\leq 0.1 + R \; \Omega$$

Where $0.1 \; \Omega$ represents the resistance of the appliance alone, and R represents the resistance of a protective conductor in the appliance cable. Resistance values below or equal to this value will pass and above this value will fail.

As you can see from Figure 9.5, the $0.1 \; \Omega$ limit is a convention that we use as the worst case value for the internal earth resistance of a

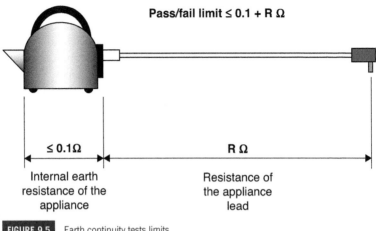

Pass/fail limit ≤ 0.1 + R Ω

≤ 0.1Ω

Internal earth
resistance of the
appliance

R Ω

Resistance of
the appliance
lead

FIGURE 9.5 Earth continuity tests limits

class I appliance when calculating the pass/fail limit. The resistance of
the appliance lead R can be calculated from the lead length and cross-
sectional area, using the same method covered earlier in this book
(assessment criteria 5.4). We then add the calculated value of R to the
0.1 Ω value, to calculate the pass/fail value for each appliance.

- *Example 1* – A class I kettle has a 0.75 mm² cable, which is 1.5 m in
  length. What is the earth continuity pass/fail limit?

A 0.75 mm² cable has a resistance per metre of 26 mΩ, or 0.026 Ω,
so a 1.5 m lead will have a resistance that is greater by 1.5 times.

$$0.026 \ \Omega \times 1.5 = 0.039 \ \Omega = R$$

We have now calculated the lead resistance R. The next step is to
add 0.1 to establish a pass/fail limit.

$$0.039 \ \Omega + 0.1 \ \Omega = 0.139 \ \Omega$$

So our pass/fail limit is ≤ **0.139 Ω**.

- *Example 2* – An extension lead has a 1.25 mm² cable, which is 10 m
  in length. What is the earth continuity pass/fail limit?

A 1.25 mm² cable has a resistance per metre of 15.6 mΩ, or 0.0156
Ω, so a 10 m lead will have a resistance that is 10 times greater.

$$0.0156 \ \Omega \times 10 = 0.156 \ \Omega = R$$

We have now calculated the lead resistance R. The next step is to add 0.1 to establish a pass/fail limit.

$$0.156 \ \Omega + 0.1 \ \Omega = 0.256 \ \Omega$$

So our pass/fail limit is $\leq$ **0.256Ω**.

- *Example 3* – **A floor scrubber has a 1.5 mm² cable, which is 15 m in length. When tested, the earth continuity result was 0.28 Ω. What would be the resistance value of the appliance alone, and is this acceptable?**

The above is a typical example of the type of question examiners may ask during your exam; they often try to mix things up a bit and get away from the standard question format. You must read this type of question carefully, all the information is there but you need to apply it in a slightly different way.

First we start out by calculating R as before. A 1.5 mm² cable has a resistance per metre of 13.3 mΩ, or 0.0133 Ω, so a 15 m lead will have a resistance that is 15 times greater.

$$0.0133 \ \Omega \times 15 = 0.1995 \ \Omega = R$$

To establish the resistance of the appliance alone, we must deduct R from the overall resistance given in the question.

0.28 Ω − 0.1995 Ω = 0.0805 Ω or **0.08 Ω** rounded down for simplicity.

So the resistance of the appliance alone is **0.08 Ω** and as this is less than 0.1 Ω it is **acceptable**.

You should practise the type of questions above as part of your exam revision until they become second nature.

So do we have to calculate the pass/fail limits for every item we test?

You will be glad to hear the answer to that question is no. For a vast majority of appliances the $\leq 0.1 \ \Omega$ limit is higher than the resistance of the appliance and the lead added together, so if we simply use this $\leq 0.1 \ \Omega$ limit as our pass/fail limit, we can save ourselves the trouble of having to calculate. The standard tests built into many PAT testers work on this principle, using 0.1 Ω as their pass/fail limit.

Unfortunately we can encounter problems with this type of equipment when testing appliances with long leads or extension leads. Their increased resistance will cause them to fail when using the 0.1 Ω limit. But when the actual limit is calculated using 0.1 + R Ω, thus allowing for the lead, the appliance will pass. When performing earth continuity testing with instruments using the 0.1 Ω limit, the tester must be very careful not to mistakenly fail items due to the incorrect test limit. In many modern test instruments it is possible to reprogram the pass/fail limits, often using a built-in limit calculator function allowing you to establish the correct pass/fail limits and use this value for future tests. For very basic PAT testers, which give only a pass/fail indication and not an actual value of the earth continuity resistance, appliances with long leads that fail the continuity test may need to be retested with another test instrument to confirm their actual serviceability.

Assessment criteria 5.6

Specify how to carry out insulation resistance testing

The *IET Code of Practice* covers insulation resistance testing in paragraph 15.5 starting on p. 93. The purpose of this test is to confirm that the various insulators within the plug, supply cable and appliance have sufficient electrical resistance to separate live parts from earth in class I appliances, and live parts from exposed conductive parts in class II appliances. Typically, under normal operating conditions, insulation needs to be capable of withstanding the supply voltage; it is for this reason that when testing insulation resistance it is desirable to use a suitably high voltage, often 500 V DC, which will place the insulation under a degree of electrical stress, allowing us to be confident that the insulation will perform correctly under real-world conditions.

Below are two diagrams illustrating how the insulation resistance test is carried out on typical class I and class II appliances:

A typical class I appliance

When performing an insulation resistance test on a class I appliance, the line and neutral conductors are connected together to avoid damaging the appliance and a measurement is made, using a test voltage of 500 V DC, between these live conductors joined together and earth. The

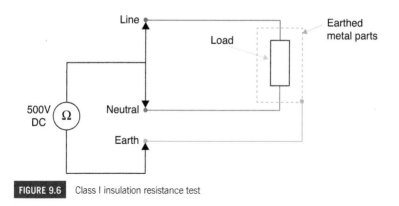

FIGURE 9.6   Class I insulation resistance test

FIGURE 9.7   Class I insulation resistance test connections example

insulation resistance for class I appliances must be at least 1 MΩ (1,000,000 Ω). For a class I appliance, the three test points, line, neutral and earth, are all connected via the three pins in the mains plug. So in this case, no additional test leads are necessary, the appliance is simply plugged into the PAT tester and the test carried out.

A typical class II appliance

When carrying out an insulation resistance test on a class II appliance, the line and neutral conductors are again connected together and the

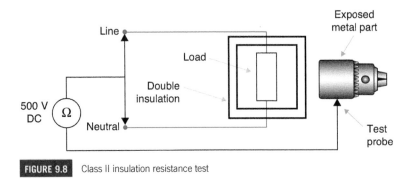

FIGURE 9.8   Class II insulation resistance test

FIGURE 9.9   Class II insulation resistance test connections example

measurement is made between these live conductors and any exposed metal parts. The insulation resistance for class II appliances must be at least 2 MΩ (2,000,000 Ω). Test connections are made using the line and neutral pins in the mains plug. A flying lead and test probe are used to make connections to any exposed conductive parts. Where the appliance has more than one exposed conductive part, the test should be repeated with the test probe connected to each part. Some appliances may have no exposed conductive parts. If this is the case then insulation resistance testing cannot be carried out, this should be recorded and brought to the attention of the responsible person.

When carrying out an insulation resistance test on class I or class II appliances, it is important to test the complete appliance, therefore the tester must confirm that the plug top fuse is intact and the appliance switch is placed in the 'on' position before testing is carried out. Some appliances containing internal relays or contactors may not fully connect the appliance until mains voltage is present. For this type of appliance the insulation resistance test may need to be supplemented by a protective conductor or touch current test, allowing the appliance's insulation to be fully tested.

Care must be taken when conducting insulation resistance tests, as exposed conductive parts may become live during the test. It is for this reason that we are recommended not to touch the appliance while the test is being carried out. To increase safety during testing, the tester may decide to take additional precautions, such as the use of insulated rubber gloves and placing the appliance on an insulating mat during testing.

Table 15.2 on p. 94 of the *IET Code of Practice* details minimum insulation resistance values. This table gives a reduced minimum insulation resistance value for heating and cooking appliances with a rating of 3 kW or higher. While it is possible for this type of equipment to yield low insulation resistance values especially while cold, great care must be taken when passing equipment with low values of insulation resistance. It is my experience that when conducting insulation resistance tests on any equipment, we should expect to see values approaching the maximum range of the tester, i.e. 99.99 MΩ on many testers. Any values below the maximum, although complying with table 15.2, should be investigated as this may indicate deterioration or a hidden defect. Typical examples of such defects would include moisture ingress or insulation damage within the appliance.

Although unlikely nowadays for most modern equipment, it is possible for some sensitive equipment to be damaged by the application of a 500 V insulation resistance test. This issue is mainly limited to class I equipment containing sensitive electronic components, but the tester must exercise caution, especially when testing expensive equipment. Where doubt exists, the test voltage should be reduced to 250 V and the reduction test voltage should be recorded. Although the test voltage is reduced, the insulation resistance pass/fail limits remain the same.

Any equipment with built-in surge protection devices (SPDs), such as surge protected extension leads and some computer power supplies, may regard the 500 V test voltage as a surge, which could activate the internal surge protection components and cause the insulation resistance reading to be lower than normal. From experience, typical values seem to be in the order of 0.3 MΩ. These tests should be repeated using a test voltage of 250 V. If the insulation resistance is still below an acceptable level, the appliance should be removed from service and guidance should be sought from the equipment manufacturer.

Assessment criteria 5.7

Specify how to carry out the protective conductor/touch current test

The *IET Code of Practice* covers protective conductor and touch current measurements in chapter 15.6 on p. 95.

There is much confusion relating to protective conductor in touch current measurements, as over the years test equipment manufacturers have referred to these tests by a variety of different names and there are also different methods used to establish these values. Generically this type of test is known as a 'leakage test' because it gives us an indication of the amount of current which has 'leaked' through the insulation. The information gained from this type of test is very similar to that of the insulation resistance test, i.e. how well is the insulation performing?

The leakage test methodology differs from the insulation resistance test by its use of the mains electrical supply (230 V or 110 V) as the test voltage. The equipment is powered up from the mains supply and the leakage test is performed while equipment is running under real-world conditions. It is for this reason that the leakage test is best able to tell us that the appliance is operating safely. Unfortunately there is a safety trade-off for this useful information. To perform this test on an as yet unproven piece of equipment, we must make it live using the mains supply voltage, which will introduce a potential electric shock risk. Also during equipment operation, rotating parts will move, heating elements will get hot etc., all of which pose a potential danger to the tester if suitable precautions are not taken. When performed correctly the leakage test can be very useful and forms an essential part of the test sequence for testing more complicated equipment.

While leakage tests can be applied to most equipment, they are typically only used where the insulation resistance test cannot fully test the equipment, usually due to the presence of a contactor or voltage operated switch, which will only activate when the mains supply is present. Where used, leakage tests should always be preceded by an insulation resistance test to reduce the chances of mains voltage being applied to faulty equipment.

As mentioned, leakage tests go by many names and there are various methods employed by the manufacturers of test equipment to establish these values. Typically leakage tests fall into two types: protective conductor current tests associated with class I equipment and touch current tests usually carried out on class II equipment.

THE PROTECTIVE CONDUCTOR CURRENT TEST

A mains supply is connected to the equipment via the portable appliance tester, the appliance is switched on and the current travelling down the class I appliance's earth wire (protective conductor) is measured. This value of current, usually displayed in mA (milliamps) is said to be the protective conductor current, i.e. the current that has leaked through the insulation to travel back to earth via the protective conductor.

The tester must observe and analyse the protective conductor current results obtained and it may be necessary to run this test for an extended period of time to fully test the appliance in all operating modes and identify any trends, i.e. that the protective conductor current is gradually

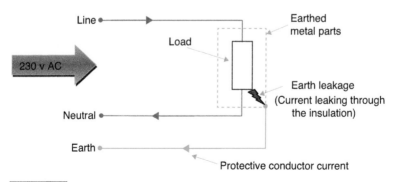

FIGURE 9.10   Protective conductor current test

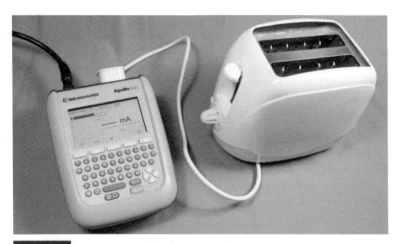

FIGURE 9.11 Protective conductor current test connections example

increasing over time, which could indicate that the insulation may fail when the appliance heats up. When interpreting these results it is also important to understand that many types of equipment, particularly those with electronic power supplies, may be designed to naturally have some leakage current travelling down the earth wire, which is not due to a fault but simply due to the power supply's normal operation. Table 15.3 on p. 96 of the *IET Code of Practice* details maximum permissible currents for protective conductor and touch current tests.

THE TOUCH CURRENT TEST

The touch current test, mainly used for class II equipment, simulates the electric shock current that would pass through a person if they were to touch an exposed conductive part of the appliance. To perform the touch current test a probe must be connected via a flying lead from the portable appliance tester to the exposed conductive part under test. Again the mains voltage is applied to the appliance via the PAT tester, the equipment is switched on and in this case a measurement is made of the current travelling back down the flying lead, this is said to be the 'touch current'.

Where the appliance has more than one exposed metal part, the test must be repeated with the test probe connected to each metal part.

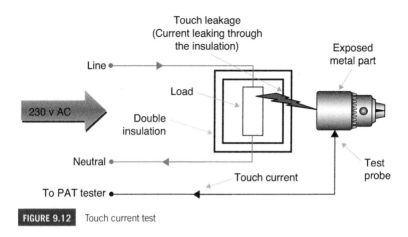

Line

Touch leakage
(Current leaking through
the insulation)

Exposed
metal part

230 v AC

Load

Double
insulation

Neutral

Touch current

Test
probe

To PAT tester

FIGURE 9.12 Touch current test

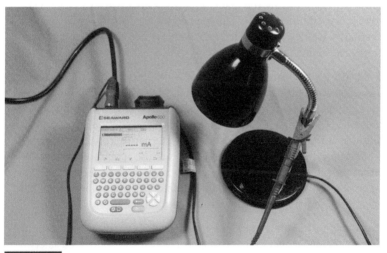

FIGURE 9.13 Touch current test connections example

In the case of Figure 9.12 we can see a common problem faced when performing a touch current test, i.e. the exposed conductive part is also a rotating part making connection of the test probe while the appliance is running impossible. Where this is the case, these parts must be tested using another, more appropriate, test method. Also remember it is not appropriate to apply the touch current test to equipment with no exposed metal parts. The touch current test can also be used to test equipment where a build-up of conductive material is suspected e.g. metal dust contamination on the case of angle grinders. The pass/fail limits of the touch current tests can be found in table 15.3 on p. 94 of the *IET Code of Practice*. Due to the construction of class II equipment, with double or reinforced insulation between live and exposed conductive parts, touch leakage currents should be virtually zero. This is reflected in the table, setting a maximum value of 0.25 mA.

SUBSTITUTE OR ALTERNATIVE LEAKAGE TESTS

The substitute or alternative leakage test is commonly available on many portable appliance testers, but most users are often confused as to the difference between this and other types of leakage test. Essentially the main difference is that the substitute or alternative leakage test is not carried out using the mains supply voltage, instead a lower test voltage is used (often 40 V). The method for conducting a substitute or alternative leakage test is exactly the same as that detailed above for the protective conductor or touch current tests.

A substitute or alternative leakage test is used where, due to the sensitivity of the equipment or the dangers posed by running the equipment, it is not appropriate to carry out either an insulation resistance test and/or a mains powered leakage test (protective conductor or touch current test). The substitute or alternative leakage test applies a low, safe voltage to the supply terminals of the appliance and measures either the protective conductor or touch currents as detailed above. The value obtained is then scaled up to estimate what the values would have been if the actual mains supply had been connected.

It is important to understand the limitations of substitute or alternative leakage testing, which must be weighed against the undoubted safety benefits for both the tester and the equipment under test. The test is carried out at a low voltage, usually 40 V, which will not place the insulation under a high degree of stress as would be the case with a 500 V insulation resistance test, so it may not detect more subtle insulation defects. Also 40 V is not sufficient to operate contactors or voltage operated switches within the equipment, thus not allowing the equipment to be fully tested. The scaling process required to simulate the test results if a mains voltage were to be applied can also reduce the accuracy as any errors are also multiplied.

The substitute or alternative leakage test is a useful test for equipment that cannot be subjected to conventional insulation resistance or leakage testing, but should be seen very much as a last resort and is not necessary were the conventional tests can be carried out. The normal pass/fail limits from table 15.3 as used for the protective conductor and touch current tests are also applicable to substitute or alternative leakage testing.

CONFIRMING THE SERVICEABILITY OF AN APPLIANCE'S INSULATION

As you can see from criteria 5.6 and 5.7, there are many different ways to confirm the serviceability of an equipment's insulation. Below is a table illustrating those options:

Table 9.3   The different methods of testing insulation

Typical test Voltage	Test method
1.5–3 kV	Dielectric strength testing (flash testing)
500 V	Insulation resistance testing
250 V	
230 V (Mains voltage)	Protective conductor current test
	Touch current test
40 V	Substitute/alternative leakage test

As can be seen, the table above has been arranged with the hardest test at the top and the softest test at the bottom. I have included dielectric strength testing for reference, however the *IET Code of Practice* recommends that this test is not carried out as part of in-service inspection and testing; it is, however, used as part of manufacturers' production testing and in some sectors, such as the hire industry. Naturally the higher the test voltage the more stress is being placed on the insulation and the more likely we are to find a defect. Unfortunately, high test voltages can also lead to equipment damage, particularly for sensitive electronic equipment, so in these cases a softer test would be preferred. The tester must decide which of the tests in the table above are appropriate and apply one or more of these to confirm the serviceability of the equipment's insulation.

FIGURE 9.14 Polarity test

Assessment criteria 5.8

Specify how to carry out polarity checks

The polarity check is only applicable to detachable appliance leads appliance leads (cord sets) and extension leads. The polarity check is used to confirm correct wiring at both ends of the lead i.e. that line connects to line, neutral to neutral, and earth to earth. Although polarity is checked visually during the formal visual inspection, it is often not possible to visually inspect the terminations at both ends of the lead, therefore a simple continuity test is carried out to confirm these connections.

Most modern portable appliance testers will have a dedicated test socket used for lead testing. One end of the lead is usually connected into the normal appliance test socket and the other end into the dedicated lead test socket. It is often necessary for adapters to be used when testing leads with a non-standard configuration at either end. The portable appliance tester will perform a wiring polarity test and indicate any fault that is found. Typical faults include an open circuit conductor, a short circuit between conductors or incorrect wiring.

For some more complicated leads or where adapters cannot be purchased, for example three phase or 110 V leads, measurements can be made using a suitable continuity tester. These tests will normally involve end-to-end continuity tests for each conductor and test between conductors to ensure no short-circuits or crossed wiring. Additionally, live tests can be performed as part of your functional checks using a simple plug-in socket tester as is used for testing mains socket outlets. Similar devices are also available for three phase and 110 V outlets; however, these live tests many not identify all polarity faults and the continuity test is preferred.

Assessment criteria 5.9

Specify how to carry out functional and load checks

A functional check is the last step before attaching the pass label and returning an appliance to service. By carrying out a visual inspection we have identified any visual faults and by performing our electrical tests we have confirmed as far as is reasonably practicable that the appliance is fit for service. However, it is possible for equipment to pass both inspection and testing, but still have hidden defects that pose a potential danger to the user. It is for this reason that we carry out functional checks before finally returning equipment to service.

When performing a functional check it is important that we observe any necessary safety precautions for the equipment's use, which could include PPE and may require special training, for example the use of equipment with abrasive wheels or welding equipment. It is for this reason that the tester may need to request the help of the user to fully test the function of more complicated equipment. Please keep in mind that during the functional check the equipment should still be considered potentially dangerous as it has not yet fully passed the full in-service inspection and test process. A functional check must fully check the operation of the equipment in all modes and the equipment may need to be run for an extended period of time to allow it to be fully tested.

Consider the function of a kettle: to boil water. So how would we perform a functional test? A small amount of water is added to the kettle, enough to cover the element. The kettle is switched on and when it boils we ensure that the thermostat operates correctly and switches off the kettle. We can also check for other problems such as leaks, and confirm the operation of the on/off switch and light. This is a basic example but should form the pattern of most functional checks. An appliance such as a washing machine generally will not need to be left running for 2 hours to complete a full load of washing, but the tester should use their judgement and test the operation in different modes, for example, fill, drain and spin, ensuring that the appliance operates safely at all times with no evidence of faults or leaks.

When performing a functional test, if any part of an appliance is found to not function correctly, the tester must consider the impact of this on the overall safety status of the item, and further investigation may need to be carried out before the item can be returned to service. Always remember if any part of the equipment fails to function correctly, even if that part is not often used by the user, it may indicate a hidden defect such as unseen damage within the equipment.

Many portable appliance testers have the ability to perform a 'load test'. A load test simply runs the appliance for a period of time and measures the power or current drawn by the appliance. Power is often displayed in kW (kilowatts) or kVA (kilovolt amps). Current is displayed in A (amps). The values obtained can be compared with the rating plate of the appliance to confirm that the appliance is operating within the stated values. Care must always be taken by the tester when comparing measured and rated values, as measured values are usually lower than those stated by the manufacturer. For example, a 650 W drill, when tested under no load conditions i.e. not drilling a material, will read significantly lower than 650 W. Any readings higher than the rated value should be investigated as this may indicate a fault with the appliance.

The load test is a common favourite with examiners and questions often catch out those not experienced in this area. A typical question, for example, would ask:

When testing a three-bar electric fire rated at 3 kW, what test would be appropriate to check the function of the three heating elements?

The answer naturally is the load test, because each bar would be rated 1 kW and if one bar was faulty we would see a load reading of 2 kW, if two bars were faulty we would see a reading of only 1 kW etc., thus allowing us to determine that all the elements are functioning correctly. This method can also be used to check the switch; i.e. switch position one equals one element and 1 kW, switch position two equals two elements and 2 kW etc.

Load testing is not normally carried out on its own; however, the load values are often displayed during the mains powered leakage tests.

Assessment criteria 5.10

Describe the requirement for testing RCDs incorporated into extension leads and multi-way adaptors

I introduced the topic of RCDs under assessment criteria 3.8, where I discussed their application and usage. Where appliances containing RCDs are used, the correct operation of the RCD must be confirmed during the functional check part of the test sequence. In addition to a visual check of the RCD and the operation of the test button, the RCD is subjected to simulated fault currents and the time it takes to disconnect the circuit is measured. These values, recorded in ms (milliseconds), are compared with the maximum values from the British or European standards to which the RCD conforms. Table 15.5 on page 99 of the *IET Code of Practice* details the maximum disconnection times for a range of RCDs commonly encountered during portable appliance testing.

All RCDs used to provide additional protection against electric shock will be rated 30 mA or less, typically all RCDs you will encounter during PAT testing will be rated 30 mA. The standard for portable RCDs is currently BS 7071, socket outlets incorporating RCDs should comply with BS 7288 and those RCDs incorporated in the fixed electrical installation will usually be designed to comply with BS EN 61008 or 61009.

When testing the operation of an RCD there are a number of different methods and tests that can be applied. When testing RCDs incorporated in the fixed electrical installation, it is common to perform a test using a simulated test current equal to half the rating of the RCD, i.e. 15 mA for a 30 mA device. Under these conditions the RCD should not operate, this test is used to confirm that the RCD is not too sensitive, which could give rise to unwanted disconnection supply when no fault is actually present. As this test is proving a negative, i.e. that the device does not operate, it is not seen as an essential safety test and therefore many portable appliance testers do not incorporate this function. However, if you are in possession of a dedicated RCD tester or electrical installation tester you may wish to perform this test to guard against unwanted nuisance tripping.

Typically RCDs are tested at the rated current (IΔn), usually 30 mA, and should disconnect within either 200 ms (BS 7071 and BS 7288) or 300 ms (BS EN 61008 and 61009). Usually values will be much lower than the pass/fail limits, typically 20 to 30 ms.

Where an RCD is providing additional protection against electric shock (rated 30 mA or less) it will also be tested at five times its rated current (5 × IΔn or 150mA) and should disconnect within 40 ms.

When performing both the IΔn and 5 × IΔn tests, it is important that they are repeated for both the positive and negative halves of the alternating current supply waveform. These tests are referred to as 0° (The positive half-cycle) and 180° (The negative half-cycle). As we are unable to determine whether a fault will occur when the supply is positive or negative, we must be sure that the RCD will operate correctly in either case. Therefore the typical test for a 30 mA RCD will include two tests (0° and 180°) at 30 mA (IΔn) and two tests (0° and 180°) at 150 mA

FIGURE 9.15   RCD testing

(5 × IΔn). The RCD must pass all tests, in addition to a functional check by pressing the test button, before it is deemed to be serviceable.

Many portable appliance testers now offer an RCD test and this is a very useful function to look out for when deciding which tester to purchase.

There are two methods commonly used for performing the RCD test: in the first method the RCD is connected directly to a mains socket outlet and the portable appliance tester is connected to the outlet on the RCD; the second option allows the RCD to be plugged into the test socket on the portable appliance tester and the test lead is connected between the RCD socket outlet and the lead test connector on the portable appliance tester.

When performing an RCD test using the first method i.e. the RCD connected directly to a socket outlet, the tester must ensure that the socket outlet is not already RCD protected, if this is not the case we run the risk of tripping the main RCD in the electrical installation during testing, possibly causing disruption to other installation users. It must also be noted that if an upstream RCD does trip while the test is being carried out we cannot be sure that the disconnection was caused by the appliance under test and therefore any disconnection times obtained will be invalid.

Some manufacturers do provide an isolating transformer, which can be inserted between the RCD and the mains socket outlet to prevent nuisance tripping upstream RCDs. For portable appliance testers that only offer RCD testing when the RCD is connected to a mains outlet, I would strongly recommend purchasing an isolating transformer to allow testing to be carried out safely. Where testing can be carried out with the RCD plugged into the portable appliance tester directly, the risk of tripping upstream devices is removed, therefore this method is strongly preferred when performing testing of portable RCDs.

Assessment criteria 5.11

Specify the requirements for testing appliance cord sets

The guidance for testing appliance lead sets is covered in chapter 15.9 on p. 97 of the *IET Code of Practice*. When it comes to in-service inspection and testing, appliance cord sets, extension leads and mains adapters are all treated as appliances in their own right. It is recommended that these appliances have their own asset numbers and are subject to testing separately from the equipment they supply.

Three-core leads are treated as class I appliances and are subjected to the tests listed below:

- Visual inspection.
- Earth continuity.
- Insulation resistance.
- Wiring polarity test.
- Functional check.

Two-core leads are treated as class II appliances and are subjected to the tests listed below:

- Visual inspection.
- Insulation resistance.
- Wiring polarity test.
- Functional check.

It is, however, very common that not all of the above tests will be applicable to many class II leads. For example, an insulation resistance test is carried out to exposed metal parts, so this test will not be possible for plastic leads. Also, many class II leads are not polarity dependent, making a polarity test unnecessary. Where no tests are applicable the lead may be subjected to visual inspection only.

Assessment criteria 5.12

Specify the particular requirements for equipment that has a high protective conductor current

This topic is covered in chapter 15.11 of the *IET Code of Practice* on page 100. Equipment with high protective conductor currents is a common topic for exam questions and often an area where people lose marks. So a little bit of concentration in this area during revision will help improve your exam performance.

When we looked at criteria 5.7, protective conductor and touch current testing, we learned that some equipment naturally has protective conductor currents, not due to a fault but due to the way in which the equipment is designed to operate. We also learned that the maximum acceptable protective conductor current in class I equipment that is not portable or hand-held is 3.5 mA. In addition to this type of equipment, which is relatively common, you may encounter equipment with protective conductor currents designed to exceed 3.5 mA, known as equipment with high protective conductor currents. This type of equipment is quite rare and specialised, but may be encountered during portable appliance testing.

If the protective conductor current of equipment is designed to exceed 3.5 mA, but not to exceed 10 mA, it must comply with the requirements listed below:

- Be supplied by an industrial plug and socket or permanently wired to the fixed installation.
- Have internal protective conductors of at least 1 mm^2.
- Have a label warning of high protective conductor currents and the necessity of an earth connection.

Where protective conductor currents are designed to exceed 10 mA, further guidance is necessary as the wiring in the fixed installation may be affected. Guidance for this type of equipment can be found in regulation 543.7 of *BS 7671* (the 17th edition wiring regulations).

Questions relating to this topic usually start with a hint to let you know that the equipment has a high protective conductor current, for example, 'An appliance has a protective conductor current of 5.6 mA'. It is important you pick up on the hint and recognise that 5.6 mA is above 3.5 mA so the equipment is classified as having a high protective conductor current and the requirements of chapter 15.11 apply.

The second part of the question will usually then refer to one of the three conditions listed above, for example, the correct wording of the warning label. Another type of question often used relates to equipment having a protective conductor current over 10 mA, for example:

An appliance has a protective conductor current of 13 mA, where would you find guidance relating to this type of equipment?

The standard answer to this type of question is regulation 543.7 in *BS 7671*.

Assessment criteria 5.13

Interpret results that come from testing

Although not covered specifically in the *IET Code of Practice*, the ability to understand and interpret test results is a key part of the competence needed when performing in-service inspection and testing. Whenever we perform a test we must know in advance what value we are expecting, what are the pass/fail limits and also, where possible, previous test results. Test results tell us much more than simply whether the value falls within preset limits, it is possible to gauge deterioration by gradual trends seen in results over time, i.e. results gradually creeping up or down. We can also see how close results are to a limit, indicating that the equipment may not stay within the required parameters during the period until it is next tested.

The interpretation of test results is often an area examiners use to test our understanding during exams. Typical questions will normally involve describing two values: one the current test value and one the previous test value. The candidate will then be asked to determine whether the equipment is deteriorating, staying the same or improving. Naturally these judgements depend on the quantity being measured, an increase in earth continuity is a deterioration whereas an increase in insulation resistance is an improvement. It is important during revision that you spend a little time familiarising yourself with this type of question as the wording is often quite tricky and can lead to mistakes.

When testing for real, the tester must be careful not to overly rely on the PAT tester to make pass/fail decisions with regards to the results obtained. Although the preset limits programmed into the PAT tester are a help, we must always look at the actual results and ensure we agree with the outcome presented to us.

CHAPTER 10

Learning Outcome 6

Learning outcome 6

Understand the information contained in documentation for in-service inspection and testing of electrical equipment

The legal duty placed on us by the *Electricity at Work Regulations 1989* requires us to ensure that equipment is maintained in a safe condition. No reference is made to documenting this process and the *Electricity at Work Regulations* does not require us to document any inspection and testing carried out. Although not a legal requirement, documenting in-service inspection and testing is an essential part of the process. Without documentation it would be very hard to prove due diligence, perform risk assessments, manage inspection and testing frequencies, and identify patterns and trends in test results. It is for these reasons that full and proper records of all inspection and testing activities should be established and maintained.

Assessment criteria 6.1

Explain the purpose of in-service inspection and testing documentation

The *IET Code of Practice* addresses documentation in chapter 8.3 on p. 56. As I mentioned above, documentation is an important part of in-service inspection and testing as it provides us with evidence to prove that inspection and testing were actually carried out and records data that can be used to inform risk assessments, make judgements with regards to frequency and enable us to identify patterns and trends.

The *IET Code of Practice* recommends that the following documents are established and maintained:

- An equipment register, listing all items within the organisation that require inspection and testing.
- An inspection and test record, detailing inspection and testing results for each item.
- A repair register, to record repairs carried out on equipment.
- A faulty equipment register, used to record equipment that is reported as faulty by the users.
- A test instrument record, to record the results of tests to prove ongoing accuracy of test instruments.

Sample forms are included in appendix V of the *IET Code of Practice* and you should familiarise yourself with them as part of your exam revision, as some questions may relate to the content of these forms or their use.

The IET sample forms are a good starting point when first establishing your inspection and test records, most organisations will use their own versions of these forms adapted to suit the specific needs of the organisation. If you have a downloading PAT tester you will more than likely use the software provided by the test equipment manufacturer, such as Seaward's PATGuard 3, which allows you to download records directly from the PAT tester, perform administration tasks on the records and produce a variety of reports automatically. Whether your records are kept electronically or on paper you must always take suitable steps with regards to security, and backing up data at regular intervals is strongly recommended.

Assessment criteria 6.2

Specify the actions to be taken with damaged or faulty equipment

This criteria relates to paragraph 8.5 in the *IET Code of Practice* on p. 57. Equipment found to be damaged or faulty during inspection and test may pose a potential danger to the tester, the user and the premises in which it is used. For this reason, damaged or faulty equipment must be removed from service without delay.

It is very important that equipment which has been removed from service is not inadvertently returned to use, so suitable precautions should be taken to quarantine it until a responsible person is able to make a decision with regard to the future of this equipment, i.e. whether it is to be repaired, scrapped or replaced. Faulty equipment should be labelled to warn users of the potential defect and removed to a safe area. It is common practice to cut the plug off faulty equipment to prevent its reuse, I strongly advise against this practice as it could give rise to danger if the user attempts to refit the plug and will certainly invalidate the warranty if the item is to be repaired.

To improve the overall safety within the organisation, whenever an item of equipment is found to be faulty the responsible person should perform an investigation to establish the root cause of the equipment's failure, i.e. what caused the fault? If we simply replace a piece of equipment with the exact same item, logic suggests that the fault will happen again when it could have been prevented. If equipment fails because it is at the end of its working life then simple replacement may be appropriate, but the lifespan of the equipment should be factored into any risk assessments performed and the inspection and test intervals adjusted accordingly. If a piece of equipment fails due to damage, the responsible person must assess whether the equipment is actually suitable for the task it is used to carry out and the environment in which it is used. Replacement with more suitable equipment may avoid this happening again in the future and therefore improve the safety for future users. Identifying patterns and trends and performing root cause analysis is one of the most important jobs undertaken by the responsible person and, if done well, will make a large contribution to improving the electrical safety within the organisation.

Assessment criteria 6.3

State why records should be kept throughout the lifetime of the equipment

The memorandum of guidance on the *Electricity at Work Regulations 1989 (HSR25)* recommends that records of maintenance including tests should be kept throughout the working life of equipment. We should therefore aim, when possible, to keep full records for all equipment from the time it was purchased to the point when it is finally removed from service and scrapped. These records may include initial purchase information such as receipts and guarantees, manufacturers' instructions, records of in-service inspection and testing, and records of any repairs during the life of the equipment. In a perfect world we would have all of this information for every piece of equipment within our organisation, so that if one day the user is injured by any piece of equipment we can use these records to prove due diligence. Most organisations, however, do not keep this level of records for all equipment, but it should be something that we aspire to as part of continuous improvement within our electrical safety systems.

Complete records also give us the ability to better perform inspection and testing, assess risk and identify patterns and trends. Manufacturers' data and records of initial testing, carried out when the equipment was new, are important benchmarks when performing ongoing inspection and testing, allowing us to better identify deterioration or changes made during the life of the equipment, i.e. when an incorrect fuse is fitted or changes in cable cross-sectional area when a flex is replaced.

CHAPTER 11

Learning Outcome 7

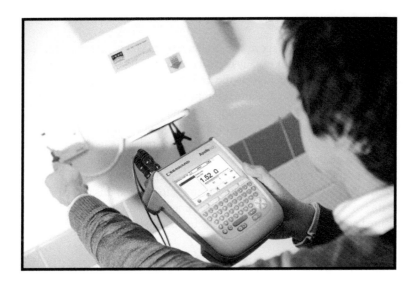

Learning outcome 7

Be able to inspect and test items of electrical equipment

This learning outcome relates directly to the practical skill of portable appliance testing, illustrated by the term 'be able' used above. I mentioned earlier that competence is a combination of knowledge and experience; this learning outcome requires that we put these things into practice by being able to perform a practical task. In order for us to be able to perform inspection and testing on items of equipment we must be confident in completing the documentation, performing a visual inspection and carrying out the necessary tests using a portable appliance tester. It is essential that we practise these tasks on a variety of equipment, both class I and class II, until they become second nature, thus building up experience as well as knowledge and improving our level of competence.

Assessment criteria 7.1

Inspect and test items of hand-held or portable electrical equipment to current industry standards
- Class I.
- Class II incorporating unearthed metal.

Your ability to meet the above assessment criteria will be tested during the practical assessment usually carried out at the end of your formal course and before you undertake the multiple-choice exam. While you need to be capable of testing any equipment you may come across, your revision should concentrate on hand-held and portable equipment, which could be class I equipment or class II equipment with exposed metal parts. Familiarise yourself with the tests applicable to this type of equipment and make sure you can correctly operate the test equipment that will be used during your assessment. Generally when it comes to practical tasks, practice makes perfect, so try to practise this type of testing over and over again until you become confident. A confident and assured performance will go a long way to demonstrate your competence to your assessor.

Assessment criteria 7.2

Complete appropriate documentation:
- Equipment register (form V.1).
- Equipment formal visual and combined inspection and test record (form V.2).
- Label (form V.3).

During your practical assessment you will also be asked to complete the documentation above, for the items outlined in criteria 7.1. Samples of these forms are included in the *IET Code of Practice*, appendix V.

In criteria 4.4, I talked about the completion of the formal visual and combined inspection test record, which forms the majority of the documentation that will need to be completed during your assessment. It is, however, important that you are familiar with completing all of the above forms and again I would recommend practising these forms during your revision. Take care when completing the forms to ensure that all boxes are completed correctly and that no boxes are left blank.

SUMMARY

In this part of the book I have covered the seven learning outcomes and their associated learning criteria. These outcomes will form the structure for all formal training courses and your exam so it is essential that you use these as the basis for your exam revision. You should also focus on being able to find these topics in the *IET Code of Practice*. Your exam is open book, so being able to find these topics easily will reduce the amount of information that you need to remember and speed up your answers during the exam. Work on each learning outcome step-by-step and you will gradually see your knowledge and competence grow.

In the next section this book I am going to look at good exam practice and sample exam questions. Use these alongside the learning outcomes as part of your exam revision.

CHAPTER 12

Good Exam Practice

The information contained in this section relates to the City and Guilds exam. (While believed to be very similar, no publicly available information could be found relating to the EAL exam, so I am unable to provide specific content relating to it.)

Your exam will be multiple-choice and computer-based. The City and Guilds exam (2377–601) lasts for 1 hour and 45 minutes and is based on 50 multiple-choice questions, meaning you have approximately 2 minutes per question. The pass mark is not officially quoted, but it is believed to be around 80 per cent, therefore the candidate must aim to get no more than ten questions wrong to achieve a pass.

The questions are distributed between the learning outcomes in the ratio displayed below.

Table 12.1 Distribution of exam questions in learning outcomes

Learning outcome	Content	Questions
Outcome 1	Statutory and non-statutory requirements	6
Outcome 2	Electrical units of measurement	3
Outcome 3	Equipment construction and classification	11
Outcome 4	Procedures for inspection and testing	8
Outcome 5	Combined inspection and testing	18
Outcome 6	Documentation	4
		50

As you can see from the table above, a majority of the questions come from outcomes 3 and 5, therefore any weakness in these two areas will probably result in a poor exam performance overall. The questions in your exam should come up in a learning outcome order and you should generally find groups of questions on the same topic arriving together as you progress through your exam.

Each question will be multiple-choice with four possible answers: A, B, C and D. You will be asked to select one of these options. You must take great care to carefully read the question and all four options before making your choice. Often, marks are lost by candidates simply misreading the question. It may be necessary to reread a question several times before deciding on your final answer. If you are still unsure, the online examination system contains a system for 'flagging' questions that you want to return to later, so always select the answer you think is correct and then 'flag' the question to remind you to look again at it later if you have time left at the end.

EXAM PREPARATION

In addition to revising the learning outcomes and sample exam questions there are some simple things you can do to help you out in the exam.

First, if you are not comfortable using a computer get some practice at home using the computer of a friend or family member. Practise using the mouse and keyboard and become comfortable with navigating your way around simple menus. A little practice at home will go a long way to alleviating the fear of using a computer in an exam scenario. Remember, the exam system has been designed to be easy to use, very little typing is necessary and an invigilator will always be on hand to help you if you get stuck. Before your exam you will also get the chance to undergo a brief tutorial on how to use the exam system and to practise the various tasks involved.

Any candidates with special requirements or learning difficulties should make these known to the examination centre as early as possible, ideally before enrolling on the course. This will give the examination centre time to make any necessary arrangements and ensure that your needs are fully met on the day. If you are not sure what help may be available, speak to your examination centre.

During your exam you are allowed to refer to a copy of the *IET Code of Practice* and it is common practice to annotate your copy prior to the exam to make things easier to find. Always confirm with your examination centre what level of annotation is acceptable to comply with their centre policy. It is common to allow highlighting or underlining of text and the use of blank page tabs to mark certain important pages. Usually centres will not allow any writing in the book or on any associated page tabs and the inclusion of any additional paperwork such as sample questions is usually forbidden. Rules relating to annotation do appear to vary from centre to centre so it is always important to confirm these prior to making any marks in your book.

If allowed to do so, you may find it useful to highlight key paragraphs relating to specific assessment criteria; highlighting of items within the contents page and index can also be useful when trying to find common topics within the book. Page tabs should be added to p. 52 (the table of initial frequencies) and p. 134 (table of resistance values). Other things you may find useful to highlight include pass/fail limits, selected definitions from part two and any relevant parts of legislation and guidance documents. When it comes to annotation, less is more as an over-annotated book can be as hard to use as one with no annotation.

In your exam you are able to use a non-programmable calculator, so ensure you take one along and that you have practised using it to perform the type of calculations that you may encounter during your exam. You are also allowed a blank sheet of paper for calculations and for making notes during the exam.

DURING THE EXAM

Once the exam begins, address each question one at a time, read them carefully (as mentioned above) and select the appropriate answer. Flag any questions you are unsure of. You are able to change your mind at any point during the exam and your answers are not set until the time ends. Displayed at the top of your exam screen will be an indication of the time remaining and your progress through the exam. Keep an eye on the time so that you can speed up or slow down as necessary, but try to leave a little time at the end to look back over any flagged questions. There are no marks for finishing early, so try to use all the time available and pace the exam so that you do not have to rush at the end.

AFTER THE EXAM

When the exam ends, leave the exam room when instructed to do so by your invigilator. Your results will usually be available within 5 or 10 minutes of finishing the exam. You will receive a result slip informing you whether you have passed or failed and giving feedback as to how you performed in each of the six learning outcomes. If you have failed the exam, look at your performance and try to identify in which areas you were weak so you are able to concentrate on those areas when revising for your retest. It is common for many candidates to fail the exam on their first try, so do not feel too despondent if you don't pass straight away. If you do fail do not rush to book a retest straightaway, allow a suitable time period for further training and revision to ensure you have a better chance next time.

CHAPTER 13

Sample Exam Questions

I have included below 50 sample exam questions, which you can use as part of your revision. You must be aware, however, that memorising these sample questions will not help you pass the exam, they are intended to give you exam practice and show you exam content and how the topics covered earlier in this book may be addressed. Your revision should concentrate on the learning outcomes and assessment criteria, which will then ensure you are able to answer the full range of questions during your exam. At the end of the section I have also included answers to the sample questions for you to refer to.

SAMPLE REVISION TEST

1. Guidance relating to the safety of test leads and probes can be found in which document?
   (a) Health & Safety Executive guidance note *GS 38*
   (b) *BS 7671*
   (c) *BS EN 60309*
   (d) Health & Safety Executive publication *PM 29*

2. The duty holders and managers of organisations should:
   (a) Be able to repair faulty electrical equipment
   (b) Know their legal responsibilities under the *Electricity at Work Regulations*
   (c) Instruct users in the use of portable appliance testers
   (d) Be competent in the use of portable appliance testers

3. The memorandum of guidance on the *Electricity at Work Regulations 1989* advises that equipment records:
   (a) Are not required for portable equipment
   (b) Should only be kept where the equipment is used in hazardous areas
   (c) Should be kept throughout the working life of the equipment
   (d) Are not required if the equipment is fed from a SELV safety supply

4. Which one of the following is not within the scope of the *IET Code of Practice*?
   (a) Fixed equipment
   (b) Portable appliances
   (c) Medical equipment
   (d) Electrical tools

5. The scope of the legislation relating to inspection and testing of electrical equipment extends from the smallest piece of equipment to distribution systems up to:
   (a) 230 V
   (b) 400 kV
   (c) 11 kV
   (d) 400 V

6. Which one of the following regulations state:
   'As may be necessary to prevent danger, all systems shall be maintained so as to prevent, so far as is reasonably practicable, such danger'?
   (a) The *Electricity at Work Regulations*
   (b) The *Management of Health and Safety at Work Regulations*
   (c) The IET wiring regulations (*BS 7671*)
   (d) The *Provision and Use of Work Equipment Regulations*

7. An insulation resistance test on a class I appliance gives a test result of 0.05M Ω. If the appliance is connected to a 230 V supply, the value of protective conductor current would be:
   (a) 46 mA
   (b) 4600 mA
   (c) 4600 A
   (d) 4.6 mA

8. If an appliance cable length is halved, the resistance value of the earth conductor will:
   (a) Remain unchanged
   (b) Halve
   (c) Increase slightly
   (d) Double

9. A value of insulation resistance recorded as 5.76MΩ is equivalent to:
   (a) 57,600 Ω
   (b) 576,000 Ω
   (c) 5,760,000 Ω
   (d) 57,600,000 Ω

10. Class II equipment may be constructed with:
    (a) Earthed metalwork separated from earthed parts by basic insulation only
    (b) Earthed metalwork separated from live parts by basic insulation only
    (c) Unearthed metalwork separated from live parts by basic insulation only
    (d) Unearthed metalwork separated from live parts by basic and supplementary insulation

11. The term 'equipment construction' relates to:
    (a) Whether the equipment is portable, movable or stationary
    (b) The equipment's suitability for use within specific hazardous environments
    (c) How the user of equipment is protected against electric shock
    (d) The degree of protection offered against the ingress of solid foreign bodies

12. To protect the user against electric shock, class III equipment relies on a supply from a:
    (a) Functional extra low voltage source
    (b) Separated extra low voltage source
    (c) Protective extra low voltage source
    (d) Reduced low voltage source

13. Where protection against electric shock is provided by using the circuit protective conductor in the fixed installation wiring, the equipment classification is:
    (a) Class 0
    (b) Class I
    (c) Class II
    (d) Class III

14. Class II equipment with a substantially continuous metal enclosure would be classified as:
    (a) Insulation-encased
    (b) Isolation-encased
    (c) Metal-encased
    (d) Metal-insulated

15. There is no provision for protective earthing or reliance on a supply from a SELV source, for which one of the following equipment?
    (a) Class I
    (b) Class 01
    (c) Class II
    (d) Class III

16. A 5 kg electric toaster is classified as:
    (a) Equipment for 'building in'
    (b) A portable appliance
    (c) Stationary equipment
    (d) A hand-held appliance

17. Which one of the following household electrical appliances may be classified as an item of stationary equipment?
    (a) A washing machine
    (b) A bathroom heater
    (c) A kettle
    (d) An electric cooker

18. A stationary appliance supplied by a flexible cable incorporating a protective conductor is classified as:
    (a) Class I
    (b) Double insulated
    (c) Metal-encased class II
    (d) Class 0

19. A refrigerator has a mass of 23 kg and is located under a kitchen worktop. The equipment type is:
    (a) Transportable
    (b) Hand-held
    (c) Portable
    (d) Stationary

20. An electric drill has basic insulation around live parts and is supplied by a three-core flexible cable. The appliance construction is classified as:
    (a) Class 0
    (b) Class I
    (c) Class II
    (d) Class III

21. User checks of stationary equipment installed for use by the public should be conducted:
    (a) Before use
    (b) Daily
    (c) Weekly
    (d) Monthly

22. During a formal visual inspection it should be confirmed that the equipment is being operated:
    (a) By a suitably trained user
    (b) In accordance with the manufacturer's instructions
    (c) By a competent person
    (d) By an instructed person

23. The suggested frequency for user checks for children's rides in an amusement arcade is:
    (a) Daily
    (b) Weekly
    (c) Monthly
    (d) Every 2 months

24. Fixed equipment in an office should be subjected to initial user checks:
    (a) Before use
    (b) Daily
    (c) Weekly
    (d) Every 3 months

25. Which of the equipment listed below is considered to be of the lowest risk?
    (a) Class I equipment in a shop
    (b) Class II equipment in an office
    (c) Class I equipment in a commercial kitchen
    (d) Class II equipment on a construction site

26. An initial formal visual inspection on a class I kettle used by guests in a hotel room should be carried out:
    (a) Weekly
    (b) Monthly
    (c) Every 12 months
    (d) Every 18 months

27. A 230 V appliance has a stated rating of 650 W. What is the correct fuse size to be fitted inside the three-pin plug?
    (a) 1 A
    (b) 3 A
    (c) 5 A
    (d) 13 A

28. The cable anchorage inside a plug must always secure the:
    (a) Line and neutral conductors
    (b) Neutral and earth conductors
    (c) Line, neutral and earth conductors
    (d) Cable sheath

29. When conducting an earth continuity test on equipment that has parts which are earthed only for functional/screening purposes, these parts should be:
    (a) Subjected to a touch current test
    (b) Tested at a current 1.5 times the fuse rating
    (c) Tested at a current between 20 mA to 200 mA
    (d) Subjected to visual inspection only

30. Before to applying an insulation resistance test to a class I electric heater, the:
    (a) Switches should be in the OFF position and all covers in place
    (b) Fuses should be checked, switches in the ON position and covers in place
    (c) Switches should be in the ON position and all covers removed if possible
    (d) Fuses should be checked, switches in the OFF position and all covers removed if possible

31. Equipment with a protective conductor current designed to exceed 3.5 mA shall:
    (a) Have a label permanently fixed indicating the location of the isolator
    (b) Have internal protective conductors with a cross-sectional area of not less than 1.5 mm²
    (c) Be permanently wired to the fixed installation or supplied by a plug and socket to *BS EN 60309*
    (d) Not be used in schools or colleges

32. The insulation of a class I domestic iron is to be tested using the protective conductor current test method. The maximum acceptable value is:
    (a) 0.25 mA
    (b) 0.55 mA
    (c) 0.75 mA
    (d) 3.5 mA

33. Which voltage should typically be applied when conducting an insulation resistance test on an electrical appliance?
    (a) 230 V AC
    (b) 500 V AC
    (c) 230 V DC
    (d) 500 V DC

34. Earth continuity testing may be carried out in certain circumstances by means of:
    (a) A low-resistance ohm meter
    (b) An insulation resistance tester
    (c) a buzzer and battery
    (d) an instrument complying with *BS 1363*

35. When testing a piece of class I equipment the first electrical test to be applied is:
    (a) Insulation resistance
    (b) Load test
    (c) aEarth continuity
    (d) Polarity

36. If an item of equipment has a high protective conductor current that is designed to exceed 10 mA, further information on the necessary precautions for such equipment can be found in:
   (a) *HSR25*
   (b) *BS 7671*
   (c) *BS 7176*
   (d) *PM 29*

37. An extension lead has a cross-sectional area of 1.25 mm² and a length of 12 m. The pass/fail limit for an earth continuity of this appliance should be:
   (a) 15.6m Ω
   (b) 187.2m Ω
   (c) 0.2872 Ω
   (d) 187.2 Ω

38. The insulation of an appliance may be deteriorating if the result of the previous insulation resistance test was, when compared with current result:
   (a) A bit lower
   (b) A lot lower
   (c) The same
   (d) A lot higher

39. An appliance is supplied with a 1.0 mm² cable, which is 2 m in length. An earth continuity test produces a reading of 0.149 Ω. What is the earth continuity for the appliance alone?
   (a) 0.039 Ω and acceptable
   (b) 0.09 Ω and unacceptable
   (c) 0.1 Ω and acceptable
   (d) 0.11 Ω and unacceptable

40. If an item of class II equipment is tested using the touch current test, the test voltage used is:
   (a) Twice the supply voltage
   (b) The supply voltage
   (c) 500 V AC
   (d) 500 V DC

41. A test for polarity is required to be carried out on appliance lead sets and:
    (a) Class I portable appliances
    (b) Class II portable appliances
    (c) Extension leads
    (d) Class III power supplies

42. An earth continuity test on a 1.5 mm², 15 m extension lead would produce a result of approximately:
    (a) 0.2 Ω
    (b) 0.1 Ω
    (c) 0.5 Ω
    (d) 1.0 Ω

43. An earth continuity test on a class I vacuum cleaner produced a result of 0.08 Ω. The same test on the same item of equipment was carried out 12 months later and produced a result of 0.57 Ω. This would indicate that the equipment has:
    (a) A better earth connection
    (b) A higher protective conductor current
    (c) Low insulation resistance
    (d) A worse earth connection

44. The measured touch current readings for a class II hair dryer must not exceed:
    (a) 0.25m A
    (b) 0.25 A
    (c) 3.5m A
    (d) 0.75 A

45. A suitable test to determine whether an electric heater with multiple elements functions correctly is:
    (a) An insulation resistance test
    (b) A load test
    (c) An earth continuity test
    (d) A touch current test

46. A measurement of earth continuity for class I equipment with a supply cable should not exceed:
    (a) 0.1 Ω
    (b) 0.1 Ω + R
    (c) 1.0 Ω + Ω
    (d) 1.0 Ω + R

47. Which one of the following items of information should not be marked on the inspection and test label?
    (a) An indication of the current safety status
    (b) The date on which the last test took place
    (c) A unique ID number
    (d) The date for retesting

48. Any equipment which fails an inspection and test shall:
    (a) Have a fail label attached and remain in use
    (b) Have a pass label attached and be withdrawn from use
    (c) Have a fail label attached and be withdrawn from use
    (d) Have a pass label attached and remain in use

49. Which of the model forms does not require the equipment serial number to be recorded?
    (a) The equipment register (form V.1)
    (b) The inspection and test record (form V.2)
    (c) The faulty equipment register (form V.5)
    (d) The test instrument record (form V.6)

50. Equipment that is faulty due to being unsuitable for the intended use should be:
    (a) Replaced with an identical new item
    (b) Replaced with a suitable item
    (c) Repaired
    (d) Tested less frequently

ANSWERS

Table 13.1   Answers to sample revision test

Question	Answer	*IET Code of Practice* references
1	A	10.1.3 (p. 63), Table IV.3, row 2 (p. 122)
2	B	9.3 (p. 60)
3	C	8.3 (p. 56)
4	C	1.2.1 (p. 20)
5	B	3.2 (p. 37)
6	A	3.1.4 (p. 34), Appendix III.1.1 (p. 115)

7	D	This is an Ohm's law question. We are required to calculate the current when given the resistance and voltage. Care must be taken in this case as the resistance is expressed in M Ω (remember 1 M Ω = 1,000,000 Ω).
		By using the Ohm's law triangle we can determine that to get the current (I) we must divide the voltage (V) by the resistance (R). V is stated as 230 V and R is stated as 0.05 MΩ, which must be converted to ohms by multiplying by 1,000,000. Therefore R equals 50,000 Ω.
		I = 230V $\div$ 50,000 Ω = 0.0046 A or 4.6 mA
		Answers A, B and C are all variations on the correct answer designed to catch out anyone who makes an error in their calculations.
8	B	This question requires you to know the relationship between the length of a conductor and its resistance. In this case halving the length of the conductor will also halve its resistance.
9	C	This question is about converting M Ω to Ω. To do this we must multiply the MΩ value by 1,000,000. Therefore 5.76 M Ω multiplied by 1,000,000 is 5,760,000 Ω
10	D	11.2 (p. 74)
11	C	11 (p. 71)
12	B	11.3 (p. 78)
13	B	11.1 (p. 72)
14	C	11.2 (p. 74)
15	C	11 This question requires a knowledge of Class 0, Class I, Class II and Class III equipment construction.
16	B	5.1 (p. 43)
17	A	5.4 (p. 44)
18	A	11.1 (p. 72)
19	D	5.4 (p. 44)
20	B	11.1 (p. 72)
21	C	Table 7.1 (p. 52)
22	B	14.1 (p. 85)
23	A	Table 7.1, Note 3 (p. 53)
24	D	Table 7.1 (p. 52)
25	B	Table 7.1 (p. 52). In this type of question, the lowest frequency equals lowest risk.

26	A	Table 7.1 (p. 52) and Note 5. A kettle is generally considered to be portable equipment.
27	B	15.13 (p. 101)
28	D	Table VII,1 row 4 (p. 135)
29	C	15.4 (p. 92)
30	B	15.5 (p. 94)
31	C	15.11 (p. 100)
32	C	Table 15.3 (p. 96)
33	D	15.5 (p. 93)
34	A	10.3 (p. 65)
35	C	15.3 (p. 91)
36	B	15.11 (p. 100)
37	C	See Table 15.1 (p. 93). The pass/fail limit for an extension lead is $0.1 + R$ Ω. From Table VI.1 on p. 134, the resistance per metre for a 1.25 mm^2 cable is 15.6m Ω/m. Therefore the resistance (R) of a 12 m is $15.6 \times 12 = 187.2$m Ω or 0.1872 Ω (remember, divide by 1,000 to convert to ohms). Then we much add on 0.1 Ω to get the pass/fail limit ($0.1 + R$ Ω). So 0.1872 Ω + 0.1 Ω = **0.2872** Ω.
38	D	When insulation deteriorates, the insulation resistance drops significantly. So we would expect to see a previous result that was a lot higher.
39	D	First we must calculate the resistance of the lead. 1.0mm^2 cable has a resistance of 19.5m Ω/m (from Table VI.1 on p. 134). So 2 metres of cable has a resistance of 19.5×2 = 39m Ω or 0.039 Ω. The measured resistance stated in the question is 0.149 Ω, so to get the resistance of the appliance alone, we must take the lead resistance away from the measured value. 0.149 Ω − 0.039 Ω = 0.11 Ω. This value is greater than the 0.1 Ω maximum allowed for an appliance alone, so therefore unacceptable.
40	B	15.6 (p. 95)
41	C	15.10.1 note 6 (p. 98)
42	A	The resistance per metre from Table VI.1 p. 134 for a 1.5 mm^2 cable is 13.3m Ω, therefore the resistance of a 15 metre cable is $13.3 \times 15 = 199.5$m Ω or 0.1995 Ω, which is approximately 0.2 Ω. Note that in this question, which does not ask for the pass/fail limits but simply the expected resistance, 0.1 Ω is not used.

43	D	If the earth continuity test has produced a higher reading than it did 12 months earlier, it has a worse earth connection. Also, we are looking for a reading of no greater than $0.1 + R \, \Omega$ so $0.57 \, \Omega$ is a very high reading, which might indicate an earth connection problem.
44	A	Table 15.3 (p. 96)
45	B	15.7 (p. 96)
46	B	Table 15.1 (p. 93)
47	D	8.4 (p. 56)
48	C	8.5 (p. 57)
49	C	Appendix V (p. 125)
50	B	Repairing an item or replacing with an identical item may lead to a repeat of the failure in future; replacement with a suitable item is the correct solution.

Suggested Further Reading

CODE OF PRACTICE

IET Code of Practice for In-service Inspection and Testing of Electrical Equipment (4th edn) ISBN 978-1-84919-626-0

HSE GUIDANCE NOTES

HS(R) 25 Memorandum of guidance on the *Electricity at Work Regulations 1989*

HS(G) 85 Electricity at Work – Safe Working Practices

HS(G) 107 Maintaining Portable Electrical Equipment

GS 38 Electrical test equipment for user on low voltage electrical systems

INDG 236 (REV3) Maintaining portable electric equipment in low-risk environments

LEGISLATION

The Health and Safety at Work etc. Act 1974 ISBN 0 10 5437743

The Electricity at Work Regulations 1989 (S.I.1989 No 635) ISBN 0 11 096635X

The Plugs and Sockets etc. (Safety) Regulations 1987 (S.I.1987 No.603) ISBN 0 11 076603

Index